BIOCHEMISTRY RESEARCH TRENDS

CYANINE DYES

STRUCTURE, USES AND PERFORMANCE

BIOCHEMISTRY RESEARCH TRENDS

Additional books and e-books in this series can be found
on Nova's website under the Series tab.

BIOCHEMISTRY RESEARCH TRENDS

CYANINE DYES

STRUCTURE, USES AND PERFORMANCE

DOUGLAS ZIMMERMAN
EDITOR

Copyright © 2019 by Nova Science Publishers, Inc.

All rights reserved. No part of this book may be reproduced, stored in a retrieval system or transmitted in any form or by any means: electronic, electrostatic, magnetic, tape, mechanical photocopying, recording or otherwise without the written permission of the Publisher.

We have partnered with Copyright Clearance Center to make it easy for you to obtain permissions to reuse content from this publication. Simply navigate to this publication's page on Nova's website and locate the "Get Permission" button below the title description. This button is linked directly to the title's permission page on copyright.com. Alternatively, you can visit copyright.com and search by title, ISBN, or ISSN.

For further questions about using the service on copyright.com, please contact:
Copyright Clearance Center
Phone: +1-(978) 750-8400   Fax: +1-(978) 750-4470   E-mail: info@copyright.com.

NOTICE TO THE READER

The Publisher has taken reasonable care in the preparation of this book, but makes no expressed or implied warranty of any kind and assumes no responsibility for any errors or omissions. No liability is assumed for incidental or consequential damages in connection with or arising out of information contained in this book. The Publisher shall not be liable for any special, consequential, or exemplary damages resulting, in whole or in part, from the readers' use of, or reliance upon, this material. Any parts of this book based on government reports are so indicated and copyright is claimed for those parts to the extent applicable to compilations of such works.

Independent verification should be sought for any data, advice or recommendations contained in this book. In addition, no responsibility is assumed by the Publisher for any injury and/or damage to persons or property arising from any methods, products, instructions, ideas or otherwise contained in this publication.

This publication is designed to provide accurate and authoritative information with regard to the subject matter covered herein. It is sold with the clear understanding that the Publisher is not engaged in rendering legal or any other professional services. If legal or any other expert assistance is required, the services of a competent person should be sought. FROM A DECLARATION OF PARTICIPANTS JOINTLY ADOPTED BY A COMMITTEE OF THE AMERICAN BAR ASSOCIATION AND A COMMITTEE OF PUBLISHERS.

Additional color graphics may be available in the e-book version of this book.

Library of Congress Cataloging-in-Publication Data

ISBN: 978-1-53616-239-4

Published by Nova Science Publishers, Inc. † New York

CONTENTS

Preface		**vii**
Chapter 1	Novel Cyanine Dyes as Inhibitors of Insulin Fibrillization *Kateryna Vus, Uliana Tarabara, Olga Zhytniakivska, Valeriya Trusova, Mykhailo Girych, Galyna Gorbenko, Atanas Kurutos, Alexey Vasilev, Nikolai Gadjev and Todor Deligeorgiev*	**1**
Chapter 2	Interactions between the Novel Cyanine Dyes and Biological Macromolecules *Olga Zhytniakivska, Kateryna Vus, Valeriya Trusova, Uliana Tarabara, Galyna Gorbenko, Atanas Kurutos, Nikolai Gadjev and Todor Deligeorgiev*	**53**
Chapter 3	Cyanine Dyes, J- and H- Aggregation in the Presence of Nanoparticles: Experimental and Theoretical Approach and Application *Dragana Vasić-Anićijević, Tamara Lazarević-Pašti and Vesna Vasić*	**123**

vi   *Contents*

Chapter 4   Cyanine Dye Aggregation within Ionic Liquid and
                  Deep Eutectic Solvent Based Systems: A Review   **193**
                  Bhawna, Divya Dhingra and Siddharth Pandey

Index   **241**

Related Nova Publications   **249**

PREFACE

Cyanine Dyes: Structure, Uses and Performance begins with evaluation of the potential of 23 novel methine cyanine dyes to inhibit the amyloid fibril formation by insulin, and the limitations in the production and storage of insulin pharmaceutical formulations.

Next, the authors describe the spectral properties of a series of monomethine, penthamethine and heptamethine cyanine dyes with an accent on their applicability to non-covalent labeling of biological macromolecules.

Additionally, the authors provide an overview of various experimental and theoretical methods for the study of cyanine dyes self-organization, as well as their applications.

The closing chapter focuses on exploring the interactions and aggregation behaviour of cyanine dyes within various ionic liquids and deep eutectic solvents-based systems.

Chapter 1 - Biomedical applications of cyanine dyes – small organic molecules, containing two nitrogen heterocyclic rings linked by a polymethine bridge, have increased dramatically in the last decade due to their excellent photophysical properties. Much attention is paid to the non-covalent labeling of proteins, nucleic acids, antibodies, with cyanines, as well as to their use as small-molecule agents for tumor-specific drug delivery, tumor photodynamic therapy, bactericidal and fungicidal compounds. However, the design of stable, non-cytotoxic, highly specific and sensitive molecular probes (including those encapsulated into the protein nanoparticles) with dual imaging and targeting properties, water- and lipid-solubility, is of current interest and is based on both the absorption/fluorescence studies and the modern quantum-chemical approaches. This investigation was aimed at the evaluation of the potential of 23 novel methine cyanine dyes to inhibit the amyloid fibril formation by insulin, associated with the development of the injection-localized amyloidosis in diabetic patients, and the limitations in production/storage of insulin pharmaceutical formulations. The trimethines AK3-3, AK3-5, and pentamethines AK5-3, AK5-9, were the most effective in suppressing the amyloid fibril nucleation and elongation under physiological conditions in vitro as monitored by the ThT fluorescence assay and electron microscopy. The increase in length and the presence of OH-groups in the aliphatic substituents on the nitrogen atom of the benzazole chromophore enhanced the dye ability to suppress the protein self-assembly. To gain deeper insight into the molecular mechanism of the modulation of insulin amyloid formation by the cyanines, the molecular docking studies were performed, showing that most mono-, tri-, penta- and heptamethine compounds interact specifically with the L17 ladder of the B chain, located at the dry interface of the amyloid cross-β spine, and form the stable complexes with the helices of the insulin monomer. The 20 ns molecular dynamics simulations of the hydrated dye-insulin monomer complexes at 37°C indicated that the total α-helical content of the protein correlates with the inhibitory potency of the examined compounds (except the heptamethine AK7-6), providing evidence for stabilization of the native insulin structure by the cyanines. The above results enabled us to suggest possible mechanisms by which cyanines can suppress the insulin amyloid formation: i) stabilization of the native protein structure followed by the retardation of the toxic oligomer formation (all compounds); and ii) blocking the lateral extension of β-sheets via the dye-protein stacking interactions (trimethines, pentamethines). Taken together, the authors' findings may prove of importance for further search for effective small molecule inhibitors of the pathological insulin aggregation for applications in medicine and biotechnology.

Chapter 2 - Cyanine dyes, photosensitive nitrogen-containing heterocyclic structures, are of special interest in the chemistry of dyes and pigments due to their extensive use in various fields of science, technology, pharmacology and biomedicine. Despite considerable progress achieved in the field of cyanine synthesis and application, the development and characterization of new fluorophores of cyanine family is of great importance. In the present study, the authors described the spectral properties of a series of monomethine, penthamethine and heptamethine cyanine dyes with an accent on their applicability to non-covalent labeling of biological macromolecules. The spectral characteristics of the novel dyes were studied in the aqueous media, in the presence of nucleic acids, proteins and model lipid membranes. The cyanine dyes under study, except of monomethines, undergo H-type self-association when free in buffer solution. Based on the results of absorption measurements, it was concluded that the examined cyanines are capable of associating with nucleic acids, proteins and lipid bilayers and this process is accompanied by the changes in the dye aggregation pattern. A tentative model for the heptamethine aggregation behavior in the presence of fibrillar and native lysozyme was formulated, assuming the dye monomer and dimer association with the protein, in addition to the dye aggregation in buffer. The association constants and stoichiometry of the dye-fibril complexation have been evaluated. The hypothesis describing the protein-dye binding mode has been proposed. The fluorescence studies showed that most dyes have negligible emission in the buffer solution as well as in the presence of DNA, proteins and model membranes, presumably due to a non-emissive nature of the dye aggregates. The exception are the monomethines exhibiting a significant emission increase upon the binding to the double-stranded DNA (dsDNA). The binding parameters for the monomethine complexation with dsDNA have been determined which are consistent with the intercalative DNA-dye binding mode. The recovered pronounced changes in the spectral responses of novel cyanines to DNA, native or fibrillar proteins and lipid bilayers allowed us to recommend these dyes for the detection and characterization of biomolecules as complementary to the existing biomarkers.

Chapter 3 - The cyanine dyes belong to the group of polymethine synthetic organic compounds which have the application in a variety of spectroscopy detection techniques in many fields of science and technology (to increase the sensitivity range of photographic emulsions, as photosensitizers in photonic devices, in biological and medicinal images to label proteins, antibodies, peptides, nucleic acid probes and any kind of other biomolecules). They undergo to the spontaneous self-organization in solution forming J- and H-aggregates, with a characteristic change of their optical properties. This self-organization of cyanine dyes is usually supported by the presence of metal ions, macromolecules or nanoparticles. The organization of dye molecules in a parallel way (plane-to-plane stacking) leads a sandwich-type arrangement (H-aggregates) with a blue-shifted absorption since a head-to-tail arrangement (end-to-end stacking) induces J-aggregates formation with a red-shifted absorption band in the absorption spectrum with respect to the monomer absorption. There is also a dramatic change in their fluorescence properties. In this paper, the authors give an overview of various experimental and theoretical methods for the study of cyanine dyes self-organization, as well as their application.

Chapter 4 - The self-aggregates of cyanine dyes owning interesting photo-physical properties have evoked great curiosity due to their remarkable technological applications in the fields, such as photography, sensors, photoconductors, medicine, and nanotechnology, among others. The highly ordered aggregates of various structures and morphologies of cyanine dyes are governed by the solvent-dye interactions. So far, while aqueous systems have been reported to be the most favorable media for cyanine dye aggregation, little has been known about dye aggregation in non-aqueous media. Ionic liquids (ILs) and deep eutectic solvents (DESs), due to their unique physicochemical properties, have shown immense potential as non-aqueous media for the molecular aggregation, though the reports are few and scarce. Recently, interesting studies have surfaced showing the dependence of cyanine dye aggregation on the identity of ILs and DESs. Thus, the unique properties displayed by ILs and DESs together with the unusual aggregation behavior of cyanine dyes make this field of research more attractive. With the aim to encourage further developments, this chapter focuses on exploring the interactions and aggregation behavior of cyanine dyes within various IL and DES-based systems.

In: Cyanine Dyes
Editor: Douglas Zimmerman

ISBN: 978-1-53616-239-4
© 2019 Nova Science Publishers, Inc.

Chapter 1

NOVEL CYANINE DYES AS INHIBITORS OF INSULIN FIBRILLIZATION

Kateryna Vus[1],*, Uliana Tarabara[1], Olga Zhytniakivska[1], Valeriya Trusova[1], Mykhailo Girych[2], Galyna Gorbenko[1], Atanas Kurutos[3], Alexey Vasilev[3], Nikolai Gadjev[3] and Todor Deligeorgiev[3]

[1]Department of Nuclear and Medical Physics,
V. N. Karazin Kharkiv National University, Kharkiv, Ukraine
[2]Department of Physics, University of Helsinki, Helsinki, Finland
[3]Faculty of Chemistry and Pharmacy,
Sofia University "St. Kliment Ohridski," Sofia, Bulgaria

ABSTRACT

Biomedical applications of cyanine dyes – small organic molecules, containing two nitrogen heterocyclic rings linked by a polymethine bridge, have increased dramatically in the last decade due to their excellent photophysical properties. Much attention is paid to the non-

---

* Corresponding Author's E-mail: kateryna_vus@yahoo.com.

covalent labeling of proteins, nucleic acids, antibodies, with cyanines, as well as to their use as small-molecule agents for tumor-specific drug delivery, tumor photodynamic therapy, bactericidal and fungicidal compounds. However, the design of stable, non-cytotoxic, highly specific and sensitive molecular probes (including those encapsulated into the protein nanoparticles) with dual imaging and targeting properties, water- and lipid-solubility, is of current interest and is based on both the absorption/fluorescence studies and the modern quantum-chemical approaches. This investigation was aimed at the evaluation of the potential of 23 novel methine cyanine dyes to inhibit the amyloid fibril formation by insulin, associated with the development of the injection-localized amyloidosis in diabetic patients, and the limitations in production/storage of insulin pharmaceutical formulations. The trimethines AK3-3, AK3-5, and pentamethines AK5-3, AK5-9, were the most effective in suppressing the amyloid fibril nucleation and elongation under physiological conditions in vitro as monitored by the ThT fluorescence assay and electron microscopy. The increase in length and the presence of OH-groups in the aliphatic substituents on the nitrogen atom of the benzazole chromophore enhanced the dye ability to suppress the protein self-assembly. To gain deeper insight into the molecular mechanism of the modulation of insulin amyloid formation by the cyanines, the molecular docking studies were performed, showing that most mono-, tri-, penta- and heptamethine compounds interact specifically with the L17 ladder of the B chain, located at the dry interface of the amyloid cross-β spine, and form the stable complexes with the helices of the insulin monomer. The 20 ns molecular dynamics simulations of the hydrated dye-insulin monomer complexes at 37°C indicated that the total α-helical content of the protein correlates with the inhibitory potency of the examined compounds (except the heptamethine AK7-6), providing evidence for stabilization of the native insulin structure by the cyanines. The above results enabled us to suggest possible mechanisms by which cyanines can suppress the insulin amyloid formation: i) stabilization of the native protein structure followed by the retardation of the toxic oligomer formation (all compounds); and ii) blocking the lateral extension of β-sheets via the dye-protein stacking interactions (trimethines, pentamethines). Taken together, our findings may prove of importance for further search for effective small molecule inhibitors of the pathological insulin aggregation for applications in medicine and biotechnology.

Keywords: insulin, amyloid fibrils, cyanine compounds, inhibitory potency, nucleation, lateral extension

INTRODUCTION

Cyanine dyes have continuously attracted tremendous interest in a variety of biomedical applications, including: i) tumor imaging and drug delivery studies (as dyes for covalent labeling of tumor-specific ligands and anti-tumor drugs) [1, 2]; ii) antitumor photodynamic therapy (as non-covalent photosensitizers) [1]; iii) detection and characterization of proteins, nucleic acids, and cells (as fluorescent markers) [3, 4, 5]; iv) pharmacology (as antibacterial and antifungal agents) [6], etc. These research activities are driven by the superior photophysical properties of cyanines, namely: i) high quantum yield in polar media such as blood serum, and strong fluorescence enhancement in the presence of specific biological targets; ii) absorption and emission bands in the near-infrared region (heptamethines); iii) feasible conjugation with various kinds of specific molecules; iv) high molar extinction coefficients. To exemplify, a number of heterocyclic heptamethines (IR-780, IR-783, MHI-148), showed preferential accumulation in cancer cells and no systemic toxicity in mice, enabling cancer imaging, and the conjugates of these dyes with chemotherapeutic agents were suitable for tumor-specific drug delivery [1]. A sensitive PicoGreen-based assay for DNA quantification was developed in the 1990th [7], followed by the enhancement of its sensitivity using the PicoGreen derivatives [8]. The monomethine cyanine dyes have been demonstrated to display a significant antibacterial activity against the strains of gram-positive bacteria [6]. Furthermore, serum albumin and β-lactoglobulin detection by capillary electrophoresis with laser-induced fluorescence was substantially improved through the implication of the squarylium compounds NN127 and SQ-3 [9].

Along with the above biomedical applications of cyanines, they also have been employed as small molecule inhibitors of pathological protein aggregation, associated with a number of human disorders, such as Alzheimer's and Parkinson's diseases, systemic amyloidosis, type II diabetes, etc. [10–12]. These disorders are featured by the accumulation of the long (>1 µM) and thin (~10–50 nm) aggregates with the cross-β-structural motif (amyloid fibrils) of specific proteins in human organs and tissues. The cytotoxic action of amyloid fibrils and their intermediates involves the formation of non-specific ionic channes, uptake of lipids into the growing fiber, cell aggregation followed by apoptosis and necrosis [13–16]. The extensive research efforts are currently focused on the search of small organic molecules capable of suppressing the amyloid fibrilization. In vitro studies have reported a huge number of such agents differing in their structure and the mode of the inhibitory effects [17–20]. Remarkably, the ability to prevent the amyloid formation has been found also for the cyanine compounds. Considerable advantages of cyanines over the other kinds of amyloid inhibitors, particularly, covalently modified peptides and DNA [17, 21, 22], include: i) the ability to easily cross the blood-brain barrier (due to the small molecular weight); ii) high stability in biological fluids and tissues; iii) the ease of chemical modification, enabling the improvements in specificity and efficiency [23–25]. Several promising anti-amyloid agents have been recently introduced, including the thiacarbocyanine N744 [26], the carbazole derivatives [10] and the 5,5'-methoxy substituted trimethine dye 7519 [27], which suppressed the tau (in vitro), Aβ (in vivo) and insulin (in vitro) pathological aggregation, respectively. However, the cyanine compounds often suffer from high cytotoxicity, and their self-association may exert non-specific inhibitory effects [27, 28]. In turn, the clinical trials have not been performed yet for most of the small molecule amyloid inhibitors, while some of them, *viz.* curcumin [18] and di-iodo form of clioquinol [20], did not show any therapeutic potency for Alzheimer's disease. Another problem is that the main attention is paid to the drugs predominantly interfering with the Aβ self-assembly, though the molecular details of their effects are not clear [22, 24, 25]. At the same time, much remains to be elucidated about the inhibitory action of small molecules on the amyloid fibrillization of the proteins other than Aβ peptide. This concerns, for instance, suppressing the insulin fibril formation, because the effective inhibitors of this process can prevent the development of injection-localized amyloidosis in diabetic patients [29], as well as facilitate the production, long-term storage, and delivery of insulin pharmaceutical formulations [30, 31].

All the above rationales unquestionably point to the necessity of designing the novel anti-amyloid agents and investigating their modes of action both in vitro and in vivo. In view of this, the present study was undertaken to assess the ability of 23 novel cyanine compounds to inhibit insulin fibrillization in vitro. To this end, we employed the fluorescence spectroscopy, electron and fluorescence microscopy, quantum-chemical calculations, molecular docking, and molecular dynamics simulations. To better understand the effects of the dye structure on its amyloid inhibitory potency, the compounds differing in the polymethine bridge length, heterocyclic substituents, the length of the aliphatic substituents on the nitrogen atom of benzazole chromophore were selected for a detailed investigation [27]. Our goals were: i) to estimate the kinetic parameters of the insulin amyloid formation in the presence of novel cyanine dyes using the thioflavin T assay; ii) to confirm the revealed inhibitory effects by the electron and fluorescence microscopy techniques; iii) to evaluate the influence of the polymethine bridge length and heterocyclic substituents on the dye ability to suppress the insulin aggregation; iv) to compare the inhibitory potencies of the cyanine and tricyanovinyl dyes; v) to establish a correlation between the quantum-chemical characteristics of the polymethine compounds and their inhibitory effects; vi) to define the possible location of the dye binding sites in native and fibrillar insulin; vii) to suggest the mechanisms by which the most effective cyanine compounds suppress the insulin amyloid formation.

MATERIALS AND METHODS

Experimental

Bovine insulin, ethylenediaminetetraacetic acid (EDTA), NaCl, dimethyl sulfoxide (DMSO), methanol, N-(2-hydroxyethyl)piperazine-N0-2-ethanesulfonic acid (Hepes), and thioflavin T (ThT) were purchased from Sigma. Monomethine [32, 33], trimethine [34], pentamethine [35] and heptamethine [36] cyanine dyes, as well as tricyanovinyl compounds

[37] (Table 1) were synthesized in the University of Sofia, Bulgaria, as described previously. Phosphotungstic acid hydrate for electron microscopy was from Reachim. The water used for this study was purified by ion exchange or distillation. All other reagents were used without further purification.

Table 1. Chemical structures of cyanine and tricyanovinyl dyes, their absorption maxima and extinction coefficients in methanol[a]/DMSO[b], the kinetic parameters of insulin fibril formation in the presence of cyanine inhibitors

Dye	Dye structures	λ_{max}	ε	F_0	F_{max}
AK12-17		507 [a]	83200 [a]	715 ± 63	8694 ± 50
AK12-18		507 [a]	77200 [a]	836 ± 28	7532 ± 21
AK12-19		517 [a]	70700 [a]	756 ± 54	6683 ± 42
AK12-20		517 [a]	59300 [a]	700 ± 48	6030 ± 76
AK3-1		628 [a]	136000 [a]	949 ± 8	3235 ± 15
AK3-3		630 [a]	159400 [a]	927 ± 5	1232 ± 6
AK3-5		631 [a]	150000 [a]	911 ± 1	1410 ± 9
AK3-7		632 [a]	161200 [a]	942 ± 4	1477 ± 10
AK3-8		652 [a]	145800 [a]	954 ± 5	4793 ± 17

Novel Cyanine Dyes as Inhibitors of Insulin Fibrillization

Dye	Dye structures	λ_{max}	ε	F_0	F_{max}
AK3-11		649 [a]	153200 [a]	954 ± 4	3367 ± 23
AK5-1		652 [b]	191158 [b]	902 ± 27	4793 ± 47
AK5-2		663 [b]	200442 [b]	899 ± 21	7634 ± 55
AK5-3		652 [b]	222041 [b]	943 ± 3	2441 ± 22
AK5-4		665 [b]	212768 [b]	939 ± 10	3523 ± 35
AK5-6		660 [b]	181648 [b]	975 ± 17	9812 ± 54
AK5-8		660 [b]	225549 [b]	858 ± 56	7590 ± 79
AK5-9		657 [b]	205688 [b]	940 ± 6	2319 ± 11
AK7-6		817 [b]	208902 [b]	905 ± 35	16839 ± 100
Cl-YO		483 [b]	72600 [b]	953 ± 25	7110 ± 55
F-YO		483 [b]	60300 [b]	905 ± 23	7138 ± 63
Cl-YO-Et		484 [b]	64000 [b]	994 ± 24	6740 ± 58
Cl-YO-Bu		485 [b]	90500 [b]	904 ± 22	2836 ± 35
YO-Pent		485 [b]	90400 [b]	879 ± 20	2034 ± 32
3b		521 [a]	46500 [a]	963 ± 33	13846 ± 98
3d		507 [a]	42300 [a]	963 ± 45	18485 ± 105

Table 1. (Continued)

Dye	Dye structures	λ_{max}	ε	F_0	F_{max}
3f		517[a]	47800[a]	1071 ± 47	13796 ± 88
3g		520[a]	79000[a]	902 ± 30	8265 ± 67
3k		503[a]	32600[a]	974 ± 29	20050 ± 124
3l		515[a]	38400[a]	1059 ± 38	21765 ± 136
3m		487[a]	32900[a]	874 ± 19	21468 ± 127
3o		504[a]	41400[a]	1033 ± 34	13937 ± 75
3q		484[a]	85900[a]	874 ± 17	21680 ± 133

Cyanine dye stock solutions were prepared in DMSO or methanol, while ThT was diluted in HEPES buffer (20 mM HEPES, 0.1 mM EDTA, 150 mM NaCl, pH 7.4). The dye concentrations were determined spectrophotometrically. The extinction coefficients (ε, M^{-1}cm^{-1}) at absorption maxima (λ_{abs}, nm) for monomethine (Cl-YO, F-YO, Cl-YO-Et, Cl-YO-Bu, YO-Pent), pentamethine (AK5-1, AK5-2, AK5-3, AK5-4, AK5-6, AK5-8, AK5-9), heptamethine (AK7-6) cyanine dyes in DMSO are presented in Table 1. The ε and λ_{abs} parameters for monomethine (AK12-17, AK12-18, AK12-19, AK12-20), trimethine cyanine dyes (AK3-1, AK3-3, AK3-5, AK3-7, AK3-8, AK3-11), and tricyanovinyl compounds (3b, 3d, 3f, 3g, 3k, 3l, 3m, 3o, 3q) in methanol can be also found in Table 1. The extinction coefficient of ThT was taken as $\varepsilon_{412} = 23800$ M^{-1}cm^{-1}.

Insulin Aggregation Assessed by Thioflavin T Assay

The insulin stock solution (10 mg/ml) was prepared in 10 mM glycine buffer (pH 1.6), followed by the protein dilution to a final concentration of

20 μM in 10 mM TRIS buffer (150 mM NaCl, 1 mM EDTA, 0.01% NaN$_3$). The kinetics of insulin fibrillization was monitored in 96-well plates filled with ThT, protein and cyanine/tricyanovinyl dye (control sample did not contain the inhibitor) loaded into the fluorescence microplate reader, heated to 37°C and incubated under constant linear shaking (50 rpm) for up to several hours (series 1) [38]. The obtained kinetic curves represent the mean of three replicates. Alternatively, orbital shaking with the increased speed (115 rpm) was also employed in the control and in the samples, containing the selected inhibitors (series 2), although the treated samples were used only for microscopy studies. The steady-state fluorescence spectra of the dyes (series 2) were recorded at 20 °C in 10 mm quartz cuvettes using Shimadzu RF-6000 spectrofluorometer and excitation and emission slit widths of 5.0 nm. In all cases (series 1 and series 2), ThT, insulin and the inhibitor concentrations in the incubated samples were 10, 20, and 10 μM, respectively. The ThT fluorescence was recorded over time at 485 nm with excitation at 430 nm.

The quantitative characteristics of the insulin amyloid formation were calculated through approximating the time (t) dependence of ThT fluorescence intensity at 485 nm (F) with the sigmoidal curve [39]:

$$F = F_0 + \frac{F_{max} - F_0}{1 + \exp[k(t_m - t)]} \tag{1}$$

where F_0 and F_{max} (Table 1) are ThT fluorescence intensities in the free form and in the presence of the protein after the saturation has been reached, respectively; k is the apparent rate constant for the fibril growth; t_m is the time needed to reach 50% of maximal fluorescence. The lag time was calculated as: $t_m - 2/k$. ThT fluorescence response to the presence of insulin fibrils was defined as F_{max} / F_0.

Transmission Electron Microscopy

For the electron microscopy assay, a 5 (10) μl drop of the tested samples, series 1 (series 2), obtained after several hours of insulin incubation under the fibril-forming conditions in the presence of inhibitors, was applied to a carbon-coated grid and blotted after 3 min (30 s). Next, a 10 μl drop of 2% uranyl acetate (1.5% (w/v) phosphotungstic acid) solution was placed on the grid, blotted after 30 s, and then washed 2 times by deionized water and air dried. The resulting grids were viewed at Tecnai 12 BioTWIN electron microscope (EM-125 electron microscope).

Fluorescence Microscopy

5 μl of the tested samples (series 2) were placed on the microscope slide and covered with a cover glass. The stained samples were viewed with Lumam-I8 fluorescence microscope.

Quantum-Chemical Calculations

The 3-21G(d,p) basis set was employed for the ground state S_0 geometry optimization of the cyanine dyes, followed by the calculation of the quantum-chemical descriptors using WinGamess (version 30 September 2017 R2) and 6-31G(d,p) basis set, as described previously [40]. Specifically, the values of the ground state dipole moment (μ_g), the energy of the ground state (E_g) and dihedral angles between the donor (quinoline moieties) and acceptor (benzazole moieties) of the dye (φ) were evaluated. Alternatively, the optimized dye conformations were further used for the estimation of other descriptors by the semiempirical PM6 method (MOPAC 2016 version 18.012L) [41]. The following characteristics were calculated: i) the energy of the highest occupied

(E_{HOMO}) and lowest unoccupied (E_{LUMO}) molecular orbitals; ii) the solvent-accessible area (CA); iii) the cosmo volume (molecular volume) (CV); iv) the molecular length (L), height (H) and width (W); v) the polarizability of the molecule at the electric field strength 0 eV (P). The Molinspiration software was used for the evaluation of the lipophilicity of the examined compounds ($LogP$), topological polar surface area ($TPSA$), and molecular weight ($Mol.wt.$) [42]. The ALOGPS method was employed to estimate the aqueous solubility ($LogS$) [43].

Docking Studies

The molecular docking studies were performed for the selected cyanine inhibitors of the insulin amyloid formation (as revealed by ThT fluorescence) in order to identify the most energetically favorable dye complexes with the native protein monomer and amyloid fibrils. The former was taken from the bovine insulin hexameric structure (PDB ID: 2ZP6), and the latter was the two-protofilament model, containing 8 strands in each protofilament, based on the crystal structure of the segment LVEALYL (the human insulin B chain residues 11–17) [40, 44]. Notably, at the first step of aggregation theinsulin dissociates into physiologically active monomers, which were therefore used in the molecular docking procedure [30, 38, 45]. The dye monomers were employed here, because these species were predominant in buffer solution at micromolar concentrations [34, 46–48], and they also showed a stronger affinity for amyloid fibrils [49], compared to dimers and higher order aggregates [50]. The top 10 energetically favorable dye-protein structures were obtained using the PatchDock algorithm that is suitable for the protein-ligand and protein-protein complexes and then refined by the FireDock, as described previously [49]. The docked complexes were viewed by the Visual Molecular Dynamics (VMD) software (version 1.9.3).

Molecular Dynamics Simulations

The 20 ns MD simulations were performed using GROMACS software (version 5.1) and the CHARMM36 force field, as described previously [40]. Briefly, the input files for MD calculations were prepared for the energetically most favorable dye-protein complexes obtained in the docking studies. Next, the topologies of the ligands were generated using the dye .mol2-files and CHARMM General Force Field and subsequently modified by replacing partial charges with those assigned by RESP ESP charge Derive Server [51]. The temperature was set at 400 K to accelerate protein unfolding that enabled us to test the stability of the free insulin monomer and the most energetically favorable dye-protein complexes. The molecules were solvated in a rectangular box and the counter ions were added to neutralize and equilibrate the system. The MD simulations of the dye-protein complexes were carried out in the NPT ensemble. The commands gmx rms, gmx rmsf, gmx gyrate and gmx sasa, included in GROMACS, were used to calculate the root-mean-square deviation ($RMSD$), the root mean square fluctuations ($RMSF$), the radius of gyration (R_g), and the solvent accessible surface area ($SASA$) per residue.

The analysis of the protein secondary structure and distances between the dye and protein centers of mass was performed in VMD, using the Tcl scripts.

RESULTS AND DISCUSSION

Kinetics of Insulin Fibrillization in the Presence of Cyanine Dyes

The process of amyloid fibril formation typically consists of two consecutive phases: a nucleation, in which partially destabilized protein monomers assemble into the oligomeric nucleus (no ThT fluorescence is observed during this phase), and a subsequent growth phase (ThT fluorescence increases until the saturation is reached) [52].

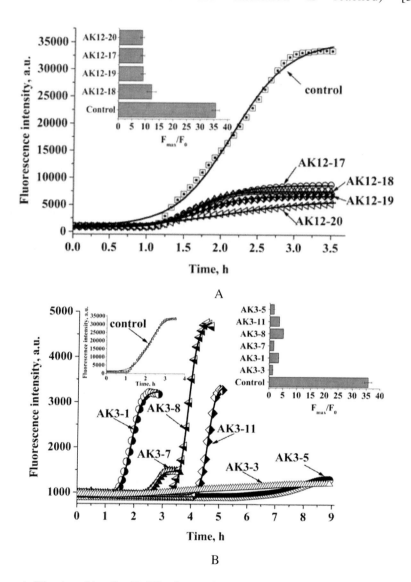

Figure 1. Kinetics of insulin fibrillization (series 1) in the presence of monomethine (A) and trimethine (B) cyanine compounds. The insets show the values of the F_{max}/F_0 parameter for each dye.

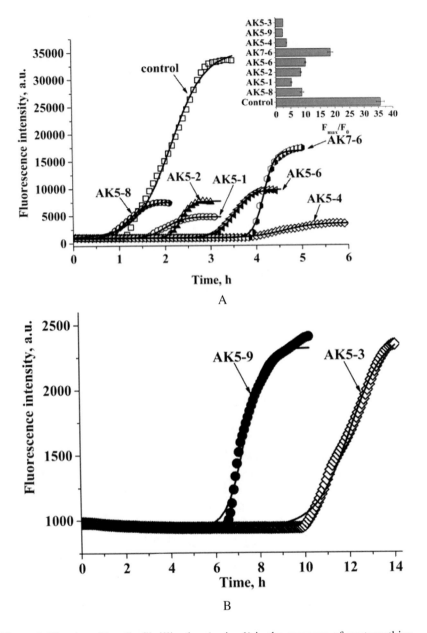

Figure 2. Kinetics of insulin fibrillization (series 1) in the presence of pentamethine cyanine compounds. The inset shows the values of the F_{max} / F_0 parameter for each dye.

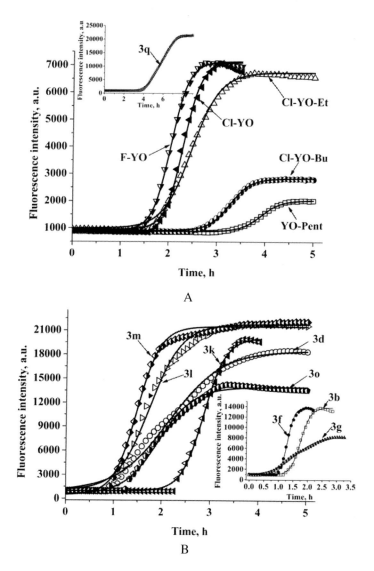

Figure 3. Kinetics of insulin fibrillization (series 1) in the presence of YO (A) and tricyanovinyl (B) compounds. The insets show the kinetic curves for the tricyanovinyl compounds 3q, 3b, 3f, 3g.

The thioflavin T kinetic curves for insulin fibrillization are shown in Figures 1–3, revealing the largest ThT fluorescence increase (F_{max}/F_0) in the control sample [53, 54]. Thus, all the tested compounds interfered with the protein aggregation and exerted inhibition effects. However, the inhibitory potency (F_0 / F_{max}) varied significantly, depending on both the polymethine bridge length and heterocyclic substituents of the dyes. Of these, the trimethine cyanines show the strongest ability to suppress the insulin aggregation, manifesting itself in 7–28 times lower F_{max} / F_0, 2–5 times lower k (AK3-3, AK3-5), and 2–5 times greater *Lag time* values than those of the control sample (Table 2). The pentamethines and monomethines (namely, five YO compounds) share the second place, displaying the 4–16-fold decrease in the ThT fluorescence response, coupled with 1.1–8 times extension of the nucleation phase compared to the control. Furthermore, 10%–40% decrease in the apparent rate constant was detected for the selected compounds including AK5-3 and AK5-9. The worst inhibitory effects on the insulin amyloid formation were observed for the heptamethine AK7-6, monomethines (AK12-17, AK12-18, AK12-19, AK12-20), and tricyanovinyl species. The presence of these dyes in the insulin sample resulted in only 1.4–4 times reduction in the F_{max} / F_0 values, 2–3 times rise in the *Lag time* (except most of the tricyanovinyl compounds), and 10%–40% decrease in the k values (for AK12-20, 3d, 3g, 3q).

The following correlations between the dye structural properties and their inhibitory potency were revealed (Table 2):

1) the long aliphatic substituents on the nitrogen atom of the benzazole chromophore enhance the effects of the monomethine (Cl-YO-Et, Cl-YO-Bu, YO-Pent compared to Cl-YO, F-YO), trimethine (AK3-3, AK3-5, AK3-7 compared to AK3-1) and pentamethine (AK5-3, AK5-4, AK5-9 compared to AK5-1, AK5-2) compounds;

2) introducing the OH-groups to the aliphatic substituents on the nitrogen atom of the benzazole moieties of AK3-7, AK5-9 induces 2–4 times higher drop in F_{max} / F_0, as compared to most other trimethines and pentamethines, although the opposite tendency was observed for the tricyanovinyl dyes (3o compared to 3k, 3l);

3) the $C \equiv N$ groups in the N-alkyl of the benzazole moiety of AK5-8, as well as in the tricyanovinyl compounds, can be responsible for a significant loss of their inhibitory activity, as compared to AK5-4 and the cyanine dyes, respectively;

4) AK3-11, AK5-6, AK7-6 showed less pronounced effects on the insulin aggregation due to the presence of benzyl group in the N-alkyl substituents of the benzazole;

5) the alkyl substituents on the quinoline ring, containing N-methylpyrrolidine and N-methylpiperidine moieties had a negative impact on the amyloid inhibitory potency of the monomethines AK12-17, AK12-18, AK12-19, AK12-20, as compared to the YO dyes;

6) at the same time, the long alkyl substituents on the quinoline ring enhanced the inhibitory effects of Cl-YO-Et, Cl-YO-Bu, YO-Pent, as compared to Cl-YO, F-YO (the F_0 / F_{max} values);

7) the replacement of S by Se in the benzazole moiety suppressed the inhibitory potency of AK5-2, AK5-4 compared to AK5-1, AK5-3, respectively;

8) a positive charge of the heterocyclic nitrogen resulted in stronger interactions and more pronounced effects of the cyanine compounds on the protein aggregation than those of tricyanovinyl dyes, although positively charged 3q induced a significant extension of the *Lag time* value;

9) surprisingly, the replacement of S by F, as well as the presence of the sulfonate or methoxy groups in the substituents on the benzazole moiety did not result in any significant variation in the inhibitory potential of the monomethines;

10) the methoxy group in the structure of 3q was responsible for the strongest inhibitory effect of this compound on insulin fibrillization, as compared to other tricyanovinyl dyes.

Table 2. Kinetic characteristics of insulin fibrillization in the presence of cyanine and tricyanovinyl compounds

Dye	k, h-1	*Lag time*, h	F_{max} / F_0
Control	2.7 ± 0.1	1.4 ± 0.1	35.7 ± 1.4
AK12-17	3.3 ± 0.1	1.1 ± 0.0	12.2 ± 1.6
AK12-18	4.8 ± 0.1	1.2 ± 0.0	9.0 ± 0.7
AK12-19	3.5 ± 0.1	1.1 ± 0.0	8.8 ± 0.7
AK12-20	1.6 ± 0.1	0.9 ± 0.0	8.6 ± 0.7
AK3-1	5.9 ± 0.2	1.6 ± 0.0	3.4 ± 0.0
AK3-3	0.6 ± 0.0	1.4 ± 0.1	1.3 ± 0.0
AK3-5	1.6 ± 0.0	6.7 ± 0.1	1.5 ± 0.0
AK3-7	7.8 ± 0.6	2.5 ± 0.0	1.6 ± 0.0
AK3-8	6.3 ± 0.1	3.6 ± 0.1	5.0 ± 0.0
AK3-11	7.9 ± 0.2	4.3 ± 0.0	3.5 ± 0.0
AK5-1	5.1 ± 0.3	1.6 ± 0.0	5.3 ± 0.2
AK5-2	8.7 ± 0.3	2.1 ± 0.0	8.5 ± 0.3
AK5-3	1.3 ± 0.0	10.5 ± 0.1	2.6 ± 0.0
AK5-4	3.1 ± 0.1	4.0 ± 0.1	3.8 ± 0.1
AK5-6	5.2 ± 0.1	3.1 ± 0.1	10.1 ± 0.2
AK5-8	5.1 ± 0.2	0.8 ± 0.0	8.8 ± 0.7
AK5-9	2.4 ± 0.1	6.5 ± 0.1	2.5 ± 0.0
AK7-6	8.6 ± 0.3	3.9 ± 0.0	18.6 ± 0.8
Cl-YO	5.4 ± 0.2	1.9 ± 0.2	7.5 ± 0.7
F-YO	5.4 ± 0.1	1.6 ± 0.1	7.9 ± 0.7
Cl-YO-Et	3.5 ± 0.1	1.9 ± 0.1	6.8 ± 0.6
Cl-YO-Bu	4.6 ± 0.1	2.9 ± 0.2	3.1 ± 0.0
YO-Pent	4.8 ± 0.1	3.5 ± 0.3	2.3 ± 0.0
3b	6.0 ± 0.2	1.5 ± 0.1	14.4 ± 1.0
3d	1.9 ± 0.1	1.1 ± 0.1	19.2 ± 1.2
3f	8.8 ± 0.3	1.1 ± 0.1	12.9 ± 0.8
3g	2.5 ± 0.1	0.8 ± 0.0	9.2 ± 0.6
3k	4.5 ± 0.1	2.5 ± 0.1	20.6 ± 1.4
3l	3.0 ± 0.1	1 ± 0.1	20.6 ± 1.3
3m	4.2 ± 0.2	1 ± 0.1	24.6 ± 1.5
3o	2.8 ± 0.1	1.2 ± 0.1	13.5 ± 0.9
3q	1.5 ± 0.0	4.4 ± 0.2	24.8 ± 1.5

The above results are in good agreement with the data obtained for the small molecule inhibitors of the pathological protein aggregation. For example, the trimethine cyanine compounds displayed the highest ability to suppress the tau aggregation [44]. Furthermore, the most effective inhibitors, AK3-3 and N744, retarded an elongation phase (k) of insulin and tau, respectively, without altering a nucleation phase and *Lag time* (Table 2) [29]. It is worth noting that the environmental factors do not always induce the changes in nucleation and elongation in the same direction [55, 56]. Next, a positive effect of the hydroxyl moieties in the inhibitor structure, observed for AK3-7, AK5-9, was also reported by Sun and coworkers for resveratrol whose OH-groups appeared to have a high affinity for amino acids [25]. In turn, the dye-protein electrostatic attraction, as well as stronger van der Waals and aromatic interactions can explain the advantages of most cyanines over the tricyanovinyl inhibitors (Table 2). Indeed, the latter are characterized by a smaller molecular volume, and a high intermolecular flexibility, that is likely to prevent the efficient dye binding to the insulin molecule [37]. The aromatic interactions of insulin with the monomethines (AK12-17, AK12-18, AK12-19, AK12-20) can also be restricted by the presence of the bulky substituents in both the benzazole and quinoline heterocycles. H-aggregation of the AK7-6 may account for the lowest potential of this compound to interfere with the insulin amyloid formation [49]. Interestingly, the pentamethines containing the long alkyl substituents in the quaternized nitrogen showed far more pronounced binding to the BSA than those with shorter chains [47]. Thus, the pentamethine and trimethine compounds seem to possess an appropriate inhibitor volume and relatively strong association with the native protein, enabling the prevention of both the amyloid nucleation and the formation of highly cytotoxic prefibrillar aggregates that is a promising approach in the design of the anti-amyloid therapeutic agents [56, 57]. This assumption is supported by the fact that the insulin hydrophobic core [38] is exposed to solvent under the fibril-forming conditions, enabling the enhanced dye-protein hydrophobic/aromatic interactions, which may prevent the transformation of the insulin α-helices into β-sheets [55].

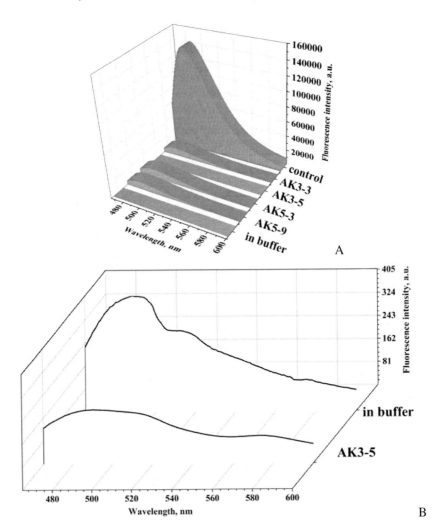

Figure 4. Fluorescence spectra of thioflavin T after insulin incubation under the fibril-forming conditions in the presence of the selected inhibitors (series 2).

The ThT fluorescence spectra measured after 115 hours of the insulin incubation under the fibril-forming conditions (series 2) in the presence of the selected trimethine and pentamethine species (Figure 4) revealed the inhibitory effects of the tested compounds, similar to those obtained in series 1, although distinct in their quantitative characteristics, that is likely due to the morphological differences between the fibrillar aggregates.

Specifically, the ThT fluorescence enhancement in the control (series 2) was about an order of magnitude higher, as compared to the dye fluorescence response in series 1 (Table 2). Such an effect may occur due to a greater extent of the fibril formation in series 2, resulting in the increased number of the accessible binding sites (the surface grooves) for the ThT molecules [58]. Alternatively, the observed results can be explained by a greater rigidity of the microenvironment of the fibril-bound ThT in series 2, resulting in a higher fluorescence quantum yield of the dye [59]. Interestingly, AK3-5 demonstrated a stronger potential to prevent the insulin amyloid formation in series 2, because the ThT did not show any fluorescence response to the protein sample, treated in the presence of the inhibitor (Figure 4B). The other compounds, AK3-3, AK5-3 and AK5-9, displayed the comparable F_{max} / F_0 values (ca. ~27), although they were an order of magnitude greater than those observed for series 1, indicating the incomplete inhibition of insulin fibrillization. These findings emphasize the importance of a careful analysis of the inhibitor potencies of small molecules at the different amyloid-forming conditions.

Notably, the above results should be treated with caution due to the following reasons:

1) the cyanine compounds can induce the morphological changes in the fibril structure, leading to a false-positive decrease in the ThT fluorescence response;

2) the absorption spectra of the monomethine and tricyanovinyl dyes overlap with those of ThT, enabling both the inhibitor excitation at 430 nm, and the Förster resonance energy transfer (FRET) between the protein-associated small molecules, resulting in a false-positive drop in F_{max} / F_0;

3) the competition between the ThT and the inhibitors for the fibril binding sites, reducing the effective amount of ThT, associated with insulin [60–62];

4) a self-association of the dye monomers on the protein scaffold into the H-dimers possessing the blue-shifted emission maxima can be

responsible for the occurrence of FRET between the ThT and the aggregated dye molecules.

In turn, a much lower turbidity of the insulin solutions incubated in the presence of cyanine dyes, compared to the control, can be regarded as evidence for inhibition of the amyloid fibril formation [63]. Furthermore, a stronger affinity and selectivity of ThT for the insulin fibrils, as compared to other dyes (Congo Red, Nile Red, etc.) [64–67], reduces the probability of the fluorophore competitive displacement from the fibril binding site. Notably, most of the tested small molecules retarded the insulin amyloid nucleation and/or elongation, showing their predominant interactions with the protein monomers. Thus, the obtained results seem to reveal the overestimated, but not the false-positive inhibitory potential of the tested compounds.

To obtain additional proofs for the ability of the novel dyes to suppress the insulin amyloid formation, and to gain deeper insight into the mechanisms contributing to this process, the fluorescence studies were complemented with the electron and fluorescence microscopy, quantum-chemical calculations, molecular docking, and molecular dynamics simulations.

Transmission and Fluorescence Microscopy Studies of the Insulin Aggregates Obtained in the Presence of the Cyanine Compounds

The transmission electron microscopy (TEM) images of the insulin fibrils formed in the absence of cyanine dyes (series 1) are given in Figure 5A, revealing the formation of the rod-like aggregates *ca.* ~200–500 nm in length and *ca.* ~20 nm in thickness similar to those observed by Iannuzzi et al. [68]. However, the extent of the protein aggregation in control was too small for the microscopic studies of the inhibition of the protein fibrillization. Therefore, to confirm the ability of the most effective cyanine dyes, viz. AK3-3, AK3-5, AK5-3, and AK5-9, to inhibit the insulin self-assembly, we performed a separate series of experiments (series 2) and viewed the obtained protein aggregates by TEM (Figure 5B–F). Indeed, these studies revealed the formation of the insulin amorphous aggregates and spherical off-pathway oligomers in the presence of the inhibitors, providing a strong evidence for the ability of the cyanine species to prevent protein fibrillization. Specifically, suprafibrillar twisted structures (*ca.* ~ 536 ± 36 nm in width and up to several tens of micrometers in length) were observed in the absence of the cyanines (Figure 5B) [69]. In turn, fuzzy amyloid-like fibrils (*ca.* ~ 27 ± 5 nm in width, and *ca.* ~ 1.2 ± 0.1 µm in length) embedded in the amorphous protein were obtained with AK3-3 (Figure 5C). Furthermore, spheroidal insulin assemblies possessing the diameters of *ca.* 30–650 nm appeared in the presence of AK3-5, AK5-3, AK5-9 (Figure 5D–F). The latter effect was similar to that reported for the insulin aggregation inhibited by eugenol, revealing the ability of the cyanines to retain native protein species growing into mature fibrils by stabilizing large oligomers [70]. Indeed, AK3-5, AK5-3, AK5-9 appeared to be the most effective in preventing the insulin amyloid nucleation (by extension of the *Lag time* value), showing the off-pathway nature the observed spheroidal assemblies. In turn, AK3-3 inhibited only the elongation of the insulin fibrils that resulted in the formation of the amyloid-containing amorphous species, being, presumably, less toxic than the large oligomers [71]. Interestingly, the observed insulin spherical aggregates (Figure 5B–F) may also represent a core of the giant insulin spherulites, containing few amounts of amyloid material, because they induced much lower ThT response than that in the control sample (Figure 4) [72]. Furthermore, secondary structure of native insulin should be partially stabilized by the most effective cyanine inhibitors due to the fact that a smallest number of spheroidal aggregates and the weakest ThT fluorescence were observed in the presence of AK3-5, as compared to that induced by AK5-3 and AK5-9 (Figure 5).

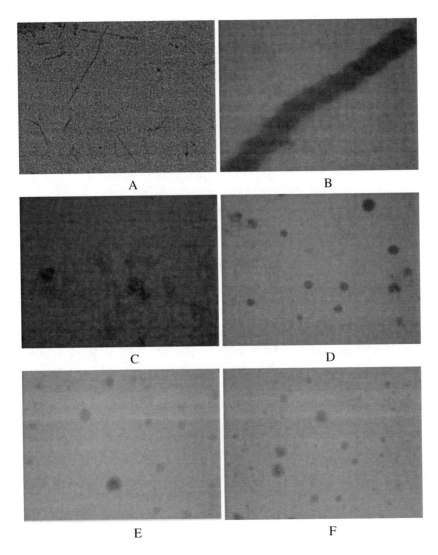

Figure 5. Transmission electron microscopy images of insulin assemblies obtained by protein incubation at 37°C, pH 7.4, 0.15 M NaCl under continuous shaking in the absence (control) or in the presence of cyanine inhibitors: A – control* (linear shaking, 50 rpm, series 1), B – control (orbital shaking, 115 rpm, series 2), C – AK3-3, D – AK3-5, E – AK5-3, F – AK5-9. Scale bars are 200 nm, 1000 nm (B), 500 nm (C, E), and 1700 nm (D, F).

Novel Cyanine Dyes as Inhibitors of Insulin Fibrillization 25

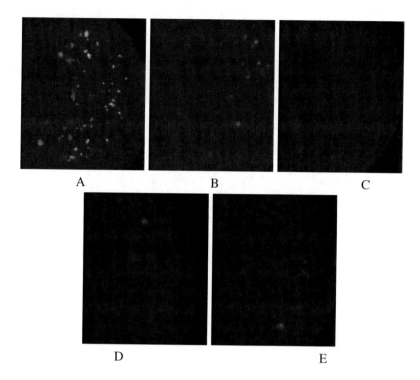

Figure 6. Fluorescence microscopy spectra of insulin suprafibrillar assemblies (series 2) in the absence (control) (A) and in the presence of cyanine inhibitors: B – AK3-3, C – AK3-5, D – AK5-3, E – AK5-9. Magnification is 300.

The samples of series 2 were also viewed by the fluorescence microscopy that revealed the insulin suprafibrillar assemblies of ~2–7 μm in diameter in control (Figure 6A). Furthermore, a smaller number of the aggregates was detected in the four inhibitory assays, containing the selected cyanine compounds (Figure 6B–E). Interestingly, fluorescence microscopy data were in agreement with the TEM results, showing no ThT fluorescence in the insulin sample containing AK3-5 (Figure 6C). Therefore, assessing the ability of the cyanine dyes to prevent pathological aggregation of various amyloidogenic proteins can be performed using the ThT fluorescence microscopy, followed by the TEM measurements of the samples containing the most promising inhibitors.

Correlation of the Inhibitory Potency with the Quantum-Chemical Characteristics of the Cyanine Dyes

The quantum-chemical calculations enabled us to establish a correlation between the experimentally determined F_{max}/F_0 values and the quantum-chemical descriptors (Tables 3–5). Remarkably, most of the examined dyes possess a planar hydrophobic core that is a common feature shared between the reported amyloid inhibitors [73, 74]. Specifically, dihedral angle φ between the benzazole and quinoline moieties of the investigated compounds varied from 163 to 180 degrees. In turn, the nonplanar monomethines (AK12-17, AK12-18, AK12-19, AK12-20) showed the lowest inhibitory potency (Table 2). Next, the $TPSA$, $LogP$, and $Mol.wt.$ values for AK3-3, AK3-5, AK5-3 were similar to those obtained for cyanine inhibitors of the tau aggregation. Of these, compound 11 capable of penetrating the blood-brain barrier appeared to be the most attractive for biological experimentations, viz. pharmacokinetics studies [26, 75, 76]. A high polarizability (P) of the planar cyanine compounds was supposed to account for the strong van der Waals interactions with a cross-β-sheet structure, followed by the stabilization of soluble tau oligomeric species [26, 77]. Furthermore, α-helices provide the surfaces suitable for the binding of π-delocalized ligands [77]. Next, a quantitative structure-activity analysis of the polyphenolic inhibitors of the Aβ aggregation showed a negative correlation between the fibrillization extent and the molecular descriptors such as P, $Mol.wt.$, CV and CA, indicating that a highest size factor leads to a greatest inhibitory effect [78]. The latter was in a good agreement with the data acquired for the YO dyes (Tables 2, 5). Indeed, Cl-YO and F-YO showed a lower ability to affect the F_{max}/F_0 and $Lag\,time$ parameters of insulin fibrillization (Table 2), as compared to Cl-YO-Bu, Cl-YO-Pent, possessing a greater L value (Table 5). In turn, the P values derived for the dyes under investigation (except the YO compounds) were lower than those reported previously. Furthermore, a negative correlation between the P, L, $Mol.wt.$ parameters and the inhibitory potency (F_0/F_{max}) was observed, with the correlation coefficients RY = -0.6, -0.75 and -0.73, respectively (Table 6). The above findings suggest that either different mechanisms are involved in the prevention of the insulin aggregation [77], or that the strong ability of the cyanines to self-associate in buffer solution and/or in the presence of native proteins prevents them from exerting specific inhibitory effects [28]. Indeed, the increase in the dye aggregation propensity with a polymethine bridge length that is proportional to the L value for all the compounds (except heptamethines) was reported previously [79]. Next, the binding of pentamethines and heptamethines to monomeric proteins turned out to shift the equilibria between various dye species towards the aggregate formation [32, 54], in contrast to the trimethines [50], producing a large decrease in the F_{max} / F_0 (Table 2).

Table 3. Quantum chemical descriptors of cyanine dyes obtained by 6-31G (d, p) basis set (GAMESS) and PM6 semi-empirical method (MOPAC)

Dye	μg, D	−Eg, Hartree	φ, deg	CA, Å2	CV, Å3	EHOMO, eV	ELUMO, eV
AK12-17	30	2263	137	506	617	-8.8	-4.0
AK12-18	35	2302	129	534	635	-9.0	-4.0
AK12-19	30	2377	130	557	658	-9.0	-4.0
AK12-20	34	2416	142	554	672	-8.9	-4.0
AK3-1	5	1352	180	374	414	-10.7	-4.4
AK3-3	6	1430	176	409	458	-10.6	-4.4
AK3-5	7	1508	176	437	503	-10.6	-4.3
AK3-7	6	1465	178	401	449	-10.7	-4.5
AK3-8	10	1657	178	464	528	-10.6	-4.4
AK3-11	11	1773	171	507	590	-10.5	-4.3
AK5-1	8	2158	165	397	443	-10.4	-4.7
AK5-2	8	6144	171	407	452	-10.3	-4.6
AK5-3	7	2236	163	419	493	-10.4	-4.6
AK5-4	8	6222	163	425	502	-10.3	-4.5
AK5-6	5	2628	176	510	633	-10.3	-4.6
AK5-8	7	2430	176	434	539	-10.5	-4.7
AK5-9	9	2385	176	425	502	-10.4	-4.6
AK7-6	1	2821	176	594	713	-10.0	-4.4

Table 4. Quantum chemical descriptors of cyanine dyes obtained by PM6 semi-empirical method (MOPAC) and the computational chemistry tools on the web

Dye	L, Å	W, Å	H, Å	P, Å3	Mol.wt.	LogP	LogS	TPSA
AK12-17	17.7	10.8	6.5	-	525	-5.3	-6.6	66
AK12-18	17.6	10.4	7.3	-	539	-5.0	-6.7	66
AK12-19	19.2	10.2	8.3	-	555	-5.3	-6.6	75
AK12-20	19.5	10.2	6.9	-	569	-5.0	-6.7	75
AK3-1	15.6	9.4	1.8	64	345	1.7	-6.13	9
AK3-3	15.6	9.4	3.0	67	374	2.6	-6.4	9
AK3-5	15.6	9.4	5.2	70	402	3.4	-6.6	9
AK3-7	15.6	9.4	3.0	66	376	1.1	-7.3	29
AK3-8	17.8	10.4	3.6	73	434	1.9	-6.4	38
AK3-11	17.9	11.5	4.8	80	480	4.1	-7.1	18
AK5-1	18.3	6.4	2.8	85	398	2.6	-6.3	9
AK5-2	18.3	6.7	4.0	80	492	2.1	-5.1	9
AK5-3	18.5	6.5	3.6	88	426	3.4	-6.5	9
AK5-4	18.6	6.4	3.5	83	520	2.9	-5.5	9
AK5-6	17.8	9.1	7.8	99	550	5.8	-7.5	9
AK5-8	18.2	6.9	3.8	89	476	1.5	-5.4	56
AK5-9	17.9	8.0	3.1	87	458	1.4	-6.1	49
AK7-6	20.8	10.2	6.7	141	616	7.1	-7.4	9

Table 5. Quantum chemical descriptors of the YO dyes obtained by PM6 semi-empirical method (MOPAC)

Dye	CA, Å2	CV, Å3	EHOMO, eV	ELUMO, eV	L, Å	W, Å	H, Å	Mol.wt.	LogP	TPSA
Cl-YO	333	373	-11.4	-4.5	12.6	8.3	4.0	323	0.8	22
F-YO	322	359	-11.4	-4.5	12.6	8.3	3.9	307	0.3	22
Cl-YO-Et	350	395	-11.3	-4.5	13.9	7.8	3.9	338	1.2	22
Cl-YO-Bu	379	438	-11.3	-4.5	14.6	7.2	5.4	366	2.2	22
YO-Pent	378	437	-11.3	-4.4	15.7	7.3	4.9	346	2.1	22

It is worth noting that a strong cross-correlation was observed between the revealed quantum-chemical descriptors, including P, L, $Mol.wt.$ (Table 6). Thus, the parameter L was selected as a best representative associated with the dye inhibitory potency (Table 6). Overall, these data indicate that despite a low self-aggregation propensity, the trimethines AK3-3, AK3-5 may have the strongest stacking interactions with the grooves on the amyloid protofibril surface, and thus, a high ability to prevent a lateral growth of the insulin fibrils [80], manifesting itself in the decrease of the k values compared to control (Table 1).

Table 6. Cross-correlation coefficients between the quantum-chemical characteristics of cyanine dyes, and correlation coefficients between the descriptors and the experimentally determined inhibitory potencies $(F_0/F_{max})^{\#}$

Dye	μg	Eg	φ	CA	CV	EHOMO	ELUMO	L	W	H	P	Mol.wt.	LogP	LogS	TPSA
μg	1	0.06	-0.93	0.55	0.54	0.92	0.85	0.23	0.44	0.62	-0.54	0.46	-0.93	-0.09	0.84
Eg		1	0.07	0.05	0.03	0.07	0.22	-0.42	0.53	-0.02	-0.21	-0.4	-0.09	-0.58	0.18
φ			1	-0.55	-0.55	-0.94	-0.78	-0.35	-0.28	-0.65	-0.13	-0.52	0.85	0.07	-0.72
CA				1	0.99	0.71	0.65	0.64	0.63	0.9	0.81	0.87	-0.3	-0.55	0.49
CV					1	0.71	0.6	0.66	0.56	0.91	0.83	0.89	-0.29	-0.52	0.52
EHOMO						1	0.81	0.46	0.4	0.76	0.9	0.68	-0.82	-0.17	0.76
ELUMO							1	0.09	0.75	0.64	-0.16	0.41	-0.75	-0.32	0.64
L								1	-0.09	0.48	0.85	0.81	-0.04	-0.02	0.26
W									1	0.5	0.00	0.24	-0.32	-0.65	0.38
H										1	0.67	0.82	-0.39	-0.49	0.49
P											1	0.85	0.78	-0.32	-0.08
Mol.wt.												1	-0.23	-0.21	0.42
LogP													1	-0.1	-0.87
LogS														1	0.00
TPSA															1
F0/Fmax#	-0.39	0.36	0.44	-0.55	-0.58	-0.55	-0.19	-0.75	-0.04	-0.51	-0.6	-0.73	0.25	-0.07	-0.38

Notably, an advantage of the tested dyes over polyphenols, e.g., curcumin and quercetin, is that their tendency to aggregate is unlikely to play a key role in the prevention of the amyloid fibril formation, and thus, the cyanines are less prone to exert non-specific inhibitory effects, which complicate the target validation [44, 81]. Next, a moderate negative correlation (RY = -0.65) between the F_0/F_{max} and HOMO-LUMO gaps indicates that the inhibitory potency of the cyanines increases along with their chemical reactivity, as well [82]. Likewise, only a weak correlation was revealed between the F_0 / F_{max} and the other descriptors (Table 6). Notably, the ground-state dipole moments of the monomethine dyes seem to be overestimated in 3-21G(d,p) basis set as judged from the typical values of this parameter reported for the cyanine compounds [83, 84]. Moreover, leaving the monomethines out of consideration did not improve the RY values. Finally, the H values allowed us to suggest that the monomethines, AK5-6 and AK7-6 are unlikely to interact with the channels running parallel to a long fibril axis because the distance between every second residue in a β-sheet is 6.5–6.95 Å [12]. Interestingly, Se has an important biological role: it is typically present in the active centers of human redox enzymes in a form of selenocysteine that is a more potent nucleophile than cysteine [85], while organoselenium molecules are of growing interest due to their important antioxidant properties [86–88]. According to our quantum-chemical calculations, the replacement of S (AK5-1, AK5-2) to Se (AK5-2, AK5-4) induced a slight increase in the dye molecular volume and solvent-accessible area, and ~15–20% decrease in its lipophilicity (Table 4). At the same time, the dye molecules retained their almost planar structures (Table 3, the torsion angle φ). These results show that hydrophobic dye-protein interactions are of great importance for the pentamethine inhibitors, leading to ~2-fold increase in the AK5-1, AK5-3 inhibitory potency (F_0 / F_{max}) as compared to that of AK5-2, AK5-4, respectively (Table 2). Interestingly, a self-aggregation of Se-containing dyes should be lower, as compared to that of S-containing species, because Se is more potent nucleophile under physiological conditions [85]. However, AK5-1 and AK5-3 monomers are likely to form more stable complexes with the hydrophobic cavities of insulin and the dry steric zipper of the insulin prefibrillar aggregates, than AK5-2 andAK5-4.

Insulin Binding Sites for the Most Effective Cyanine Inhibitors Revealed by Molecular Docking

The molecular docking studies (Figure 7A) indicated that the cyanine compounds (except AK7-6) and ThT tend to form the most energetically favorable complexes with L17 ladder (4–6 strands) of the B chain, located on the dry steric zipper of the insulin fibril protofilament, with their long axes parallel to the fibril axis, although the angle between the AK12-17 plane and the fibril extension direction was about 20 degrees [23, 40]. Furthermore, the above binding site also includes the residues C19 and L6 of the first and second protofilaments (B chain), respectively. Indeed, according to the molecular dynamics simulations, the ladders formed from Y and L possessed a high affinity for ThT [89]. The suggested binding mode was prevalent for the pentamethine AK5-3, trimethines (AK3-1, AK3-3), monomethines (the YO dyes), tricyanovinyl compounds (3g, 3q), and ThT, but not for the heptamethine AK7-6 and monomethine AK12-17, showing far less pronounced inhibitory effects than the other dyes [40]. Interestingly, despite a large thickness of some dyes (Table 4), all the compounds are capable of associating with fibrillar insulin parallel to a long fibril axis due to location of the protein binding site at the C-termini of LVEALYL β-strands (not in the grooves formed by the side chains of amino acids) (Figure 7A). In the most energetically favorable dye-protein complex AK7-6 was attached to the fibril along the β-strand direction, interacting with L17, Q4, and L6 residues of the first and second protofilament (B chain), respectively [40]. However, the non-specific primary binding sites for AK7-6, AK12-17, AK12-18, AK12-19, AK12-20 were observed, embracing the C- and N-termini of β-strands (coil structure) located at the fibril polar face [40]. Most trimethines and pentamethines also associated with the coil structure of the insulin fibril (secondary binding modes, data not shown).

The analysis of the fibril binding sites revealed that the possibility of the competitive binding of the YO, trimethine, pentamethine cyanine dyes and ThT, in principle, cannot be excluded. However, the inhibition of the amyloid growth by, for instance, resveratrol, naproxen and ibuprofen was attributed to their binding to the surface grooves on the edge of IAPP and Aβ fibrils, followed by blocking the lateral extension of the fibril [89, 90]. In our case, the structure of the dry steric zipper interface can be destabilized by the bound compounds, resulting in the prevention of insulin cross-β spine formation [38, 91–93]. The docking results coincide with the experimental data, showing that the most effective inhibitors, viz. AK3-3, AK3-5, AK5-3, and AK5-9 suppressed the fibril elongation, i.e., modulated the aggregation of the preformed nuclei. In turn, as in the case of resveratrol, no arguments were obtained in favor of the dye binding to the edge of the ending β-strand, which may inhibit a longitudual fibril growth [17, 93]. Furthermore, our data showed that the monomethine compounds (AK12-17, AK12-18, AK12-19, AK12-20), AK7-6 and ThT do not compete for the same protein binding sites, making it unlikely that ThT is displaced from the fibril side-chain channel. Rather, the ThT fluorescence intensity is reduced due to the inhibition of insulin aggregation [40]. Similarly, the Förster resonance energy transfer between ThT and monomethines (except the YO dyes) is hardly possible due to the lack of specific binding of the dyes to amyloid fibrils. In contrast, the YO dyes may compete with ThT for the same fibril binding sites (data not shown).

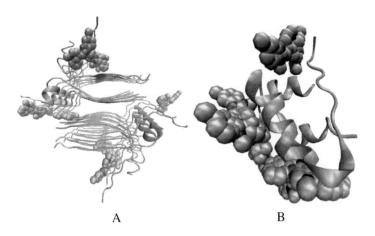

Figure 7. Representative docked poses (10 the most energetically favourable complexes) of fibrillar (A) and native (B) insulin with Cl-YO.

The most energetically favorable cyanine-protein complexes (Figure 7) were formed as a result of the dye binding to different sites of the native insulin [40]:

1) Cl-YO, AK5-3 and AK7-6 were attached to the B-chain α-helix (residues 10–17) and to the B-chain residues 1-4 (β-sheet includes the residues B3, B4);
2) AK3-1, AK12-17 and ThT associated with the B-chain residues 17–22 (3-10-helix includes the residues B20–B22), with the B-chain residues 1-4 (AK3-1), F1 (ThT), as well as with the A-chain α-helix (residues 13–17) (AK12-17);
3) AK3-3 was bound to the B-chain residues 10–13 and F1.

More specifically, the typical binding modes for cyanines are:

1) the association with the B-chain residues 1-4, the B-chain α-helix and/or 3-10-helix, that is preferable for all the dyes and may hamper a nucleus formation by preventing the interactions between the B-chains of the two partially unfolded insulin monomers [38, 93];
2) the binding between the B-chain C-terminal fragment (coil) and the A-chain α-helix (residues 2–8) that is observed for ThT and AK7-6;
3) the binding between the B-chain 3-10 helix (residues 20–22) and the A-chain α-helix (residues 13–17) that is found for AK12-17 and AK3-1;
4) the association with the B-chain C-terminal fragment (coil) and with the A-chain α-helix (residues 2–8) that is revealed for AK3-3.

On the whole, the cyanines seem to have a high affinity for α-, 3-10-helices of the B-chain and the A-chain α-helix (residues 13–17), although, e.g., AK12-17 and AK7-6 are characterized by multiple binding sites, contrary to other probes [40, 93, 94]. The latter may reflect weak binding of the monomethines (except the YO compounds) and heptamethines to the native insulin due to the lack of specific binding sites. This also can account for their significantly lower (non-specific) inhibitory effects on the insulin aggregation (Table 2) [95]. The rest of the monomethines also lacked the specific binding sites, while the pentamethines were bound to the B-chain α- and 3-10 helices, and the trimethines formed stable complexes with both B-chain and A-chain helices.

The molecular docking results support the above assumption that the novel cyanines are capable of stabilizing the native insulin structure by interacting with the B-chain α-helix (comprising the residues 11–17 of the amyloid core [38]). Likewise, Congo Red and other organic molecules were reported to inhibit insulin fibrillization [96, 97]. In turn, slowing down the insulin transformation from α-helical- to the β-sheet-rich structure was also induced by 1,2-Bis[4-(3-sulfonatopropoxyl)phenyl]-1,2-diphenylethene, possessing a stronger affinity for a partially unfolded protein conformation compared to the native one [93]. The fact that hydrophobic (e.g., leucine, valine, alanine) and aromatic (e.g., the B-chain F1 and the A-chain Y14) residues constitute the insulin binding site for the cyanine dyes, points to the key role of hydrophobic and aromatic intermolecular interactions in the stabilization of the dye-protein complexes. Indeed, fibril formation by the human insulin with mutations to more polar residues retarded the amyloid nucleation, reflecting the high impact of hydrophobic interactions on the protein aggregation [98]. At the same time, the polyphenolic inhibitors (luteolin, transilitin, maritimetin) of the Aβ aggregation showed π-π stacking interactions with the peptide [78]. Interestingly, bovine insulin is significantly more prone to fibrillization than human and porcine insulin due to the presence of the A8 residue (A-chain) on the protein surface, enhancing hydrophobic intermolecular interactions [51]. From this point of view, the ability of the examined dyes to associate with the A8 residue resulting from the fact that these compounds (except AK12-17, AK12-18, AK12-19, AK12-20, Cl-YO, F-YO) possess high *LogP* values (similar to compound 11 [44]) can lead to the increased *Lag time* value of insulin fibrillization (Table 2). However, despite the inhibitory potency of many anti-amyloid therapeutic agents is connected with their tendency to self-associate due to strong stacking interactions with aromatic amino acids, the cyanine dye ability to prevent insulin from the aggregation showed inverse correlation with the L value (except the YO compounds). These results suggest that the primary inhibition mechanism of the insulin amyloid formation is the stabilization of the α-helical structure (i.e., the retardation of the protein nucleation, as supported by the docking studies). Alternatively, the dye-protein stacking interactions (blocking the lateral extension of β-sheets) may be weaker than those within the dye assemblies, leading to a less pronounced effect on the F_{max} and k, coupled with the increase in the L values [99,100]. Finally, the trimethines occupy both A- and B-chain α-helices, a feature that may underlie their ~2–4 times higher inhibitory potency than that of the pentamethines(Table 2, F_{max} / F_0). Indeed, when the B-chain C-terminus was tethered with the A-chain N-terminus, nontoxic amorphous aggregates of human insulin were revealed [101]. It is noteworthy that electrostatic interactions may also play an essential role in the formation of a critical nucleus, as was demonstrated, in particular, for the human insulin mutants with altered charge, for which the increase in the $Lag\ time$ value was detected [94]. Therefore, the cationic cyanine dyes may inhibit the insulin nucleation by interaction with negatively charged amino acids, as well.

Molecular Dynamics Simulations of Insulin Interactions with the Most Effective Cyanine Inhibitors of the Protein Fibrillization

Finally, to verify the idea that the inhibitory effect of cyanines on insulin fibrillization may arise from the stabilization of the native protein structure, we performed a series of molecular dynamics simulations for the five energetically most stable dye-protein complexes. The helicity of free insulin averaged over all trajectory was ~32% [40]. Similarly, three α-helices of human insulin (their positions are similar to those of bovine insulin) retain about 50% of their structure at 60 °C, as revealed by CD measurements [102]. It appeared that the dyes under study are capable of stabilizing the insulin native structure [40]. As illustrated in Figure 8, the average helicity was higher in the presence of cyanines, with the magnitude of this effect increasing in the row: AK12-17 < AK3-1 < AK7-6 < AK5-3 < AK3-3.

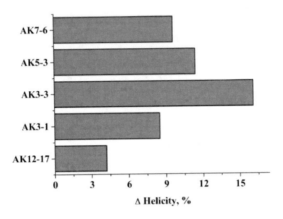

Figure 8. Changes in the insulin helicity (averaged over MD trajectory) in the presence of the selected cyanine dyes with respect to the control (without the dyes).

The backbone RMSD and radius of gyration steadily increased as a function of time indicating that the protein structure is significantly denatured during about 10 ns (Figure 9). In general, this row correlates with the inhibitory potency of the examined compounds, except AK7-6. The observed stabilization of the insulin structure under the influence of the cyanines may result from their ability to interact with the B-chain α-helix as revealed by the molecular docking. Likewise, the SASA per residue decreased in the B-chain (except AK7-6) and A-chain (except AK12-17) helical regions in the presence of cyanines, that is consistent with the involvement of hydrophobic interactions in the formation of the dye-protein complex (Figure 9) [40, 103]. Similarly, small stress molecules, the inhibitors of the insulin amyloid formation, improved the thermal stability of the protein by stabilizing the native conformation [97]. Interestingly, AK3-3, possessing the highest inhibitory potency, induced the most significant drop in the RMSF in the helical regions, compared to other dyes (Figure 9A). Thus, the reduced amino acid flexibility supports the suggestion that the cyanine compounds improve the stability of native insulin similarly to e.g., b-cyclodextrin [104].

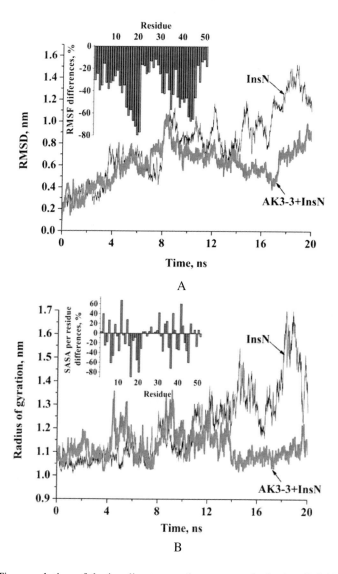

Figure 9. Time evolution of the insulin structural parameters in the insulin/ AK3-3 complex: the RMSD (A), the radius of gyration (B). The insets show RMSF differences, % (A), and SASA per residue differences, % (B).

The decreased backbone RMSD and radius of gyration (Figure 9A) provide evidence for the lower denaturation extent and higher degree of structural compactness of insulin in the presence of cyanine dyes [40]. However, the high RMSD values also suggest the possibility of the partially unfolded insulin stabilization by the investigated compounds. The latter can be explained by more favorable binding to the conformations with the solvent-exposed hydrophobic residues, as was shown, e.g., for the organic fluorogen BSPOTPE inhibiting the nucleation and elongation of the insulin amyloid fibrils at low dye-protein ratios [93].

Based on the above results, the possible inhibition mechanisms of the insulin aggregation by the cyanine dyes can be outlined as follows (Figure 10):

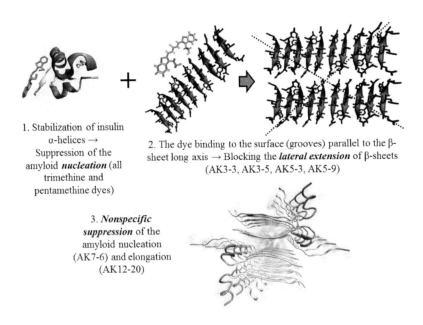

Figure 10. The proposed inhibition mechanisms of the insulin amyloid formation by the cyanine dyes. The insulin and dye molecules are shown as new cartoon and bonds, respectively.

1) stabilization of the native structure by hydrophobic, aromatic and electrostatic interactions, followed by the retardation of the protein nucleation (all dyes);

2) blocking the lateral extension of β-sheets via the dye-protein stacking interactions (AK3-3, AK3-5, AK5-3, AK5-9);
3) nonspecific suppression of the protein aggregation (AK12-20, AK7-6).

CONCLUSION

In summary, the novel cyanine compounds inhibited the insulin amyloid fibril formation under the physiological conditions in vitro, as revealed by the fluorescence and microscopy studies. The trimethines (AK3-3, AK3-5) and pentamethines (AK5-3, AK5-9) containing long aliphatic substituents on the nitrogen atom of the benzazole chromophore showed the most pronounced inhibitory potencies, including the retardation of both the amyloid nucleation and elongation. Furthermore, a strong negative correlation between the molecular length of the dyes and their ability to inhibit insulin fibrillization was uncovered by the QSAR analysis. In turn, formation of the amyloid nuclei was substantially delayed by the most hydrophobic compounds. Notably, the cyanine dyes showed greater inhibitory effects on insulin fibrillization as compared to the tricyanovinyl compounds, possessing lower molecular weight and lipophilicity. The molecular dynamics simulations of the energetically favorable dye-protein complexes obtained by the molecular docking, indicated the stabilization of the insulin α-helices by the most effective inhibitors of the pathological protein aggregation, unlike many polyphenols (viz., myricetin, baicalein, tannic acid, orcein), which do not interfere with the early nucleation events [22, 105]. The most promising cyanine inhibitors are expected to possess low cytotoxicity due to the low affinity for DNA [50], and the ability to easily cross the blood-brain barrier arising from their high lipophilicity and low molecular weight [26, 75, 76]. In this context, further evaluation of the amyloid inhibitory potency of these compounds at nanomolar concentrations, followed by the pharmacokinetic studies, may prove essential for the development of the effective and nontoxic anti-amyloid pharmaceutical formulations [26].

REFERENCES

[1] Shi, C., Wu, J. B., Pan, D. (2016). Review on near-infrared heptamethine cyanine dyes as theranostic agents for tumor imaging, targeting, and photodynamic therapy. *J. Biomed. Opt.*, 21: 50901.

[2] Bricks, J. L., Kachkovskii, A. D., Slominskii, Y. L., Gerasov, A. O., Popov, S. V. (2015). Molecular design of near infrared polymethine dyes: A review. *Dyes and Pigm.*, 121: 238–255.

[3] Patonay, G., Salon, J., Sowell, J., Strekowski, L. (2004). Noncovalent labeling of biomolecules with red and near-infrared dyes. *Molecules*, 9: 40–49.

[4] Patsenker, L., Tatarets, A., Kolosova, O., Obukhova, O., Povrozin, Y., Fedyunyayeva, I., Yermolenko, I., Terpetschnig, E. (2008). Fluorescent probes and labels for biomedical applications. *Ann. N. Y. Acad. Sci.*, 1130: 179–87.

[5] Shindy, H. A. (2017). Fundamentals in the chemistry of cyanine dyes: A review. *Dyes and Pigm.*, 145: 505–513.

[6] Abd El-Aal, R. M., Younis, M. (2004). Synthesis and antimicrobial activity of certain novel monomethine cyanine dyes. *Dyes and Pigm.*, 60: 205–214.

[7] Singer, V. L., Jones, L. J., Yue, S. T., Haugland, R. P. (1997). Characterization of PicoGreen reagent and development of a fluorescence-based solution assay for double-stranded DNA quantitation. *Anal. Biochem.*, 249: 228–38.

[8] Khatoo, M., Yang, J., Gee, K. R., Gasser, S. (2019). Evaluation of PicoGreen variants for use in microscopy and flow cytometry. *BioRxiv.*, doi: 10.1101/584037.

[9] Welder, F., Paul, B., Nakazumi, H., Yagi, S., Colyer, C. L. (2003). Symmetric and asymmetric squarylium dyes as noncovalent protein labels: a study by fluorimetry and capillary electrophoresis. *J. Chromatogr. B Analyt. Technol. Biomed. Life Sci.*, 793: 93–105.

[10] Li, Y., Xu, D., Ho, S. L., Li, H. W., Yang, R., Wong, M. S. (2016). A theranostic agent for in vivo near-infrared imaging of β-amyloid

species and inhibition of β-amyloid aggregation. *Biomaterials*, 94: 84–92.

[11] Necula, M., Chirita, C. N., Kuret, J. (2005). Cyanine dye N744 inhibits tau fibrillization by blocking filament extension: implications for the treatment of tauopathic neurodegenerative diseases. *Biochemistry*, 44: 10227–10237.

[12] Krebs, M. R., Bromley, E. H., Donald, A. M. (2005). The binding of thioflavin-T to amyloid fibrils: localisation and implications. *J. Struct. Biol.*, 149: 30–37.

[13] Caughey, B., Lansbury, P. T. (2003). Protofibrils, pores, fibrils, and neurodegeneration: separating the responsible protein aggregates from the innocent bystanders. *Annu. Rev. Neurosci.*, 26: 267–298.

[14] Sparr, E., Engel, M. F. M., Sakharov, D. V., Sprong, M., Jacobs, J., de Kruijf, B., Hoppener, J. W. M., Killian, J. A. (2004). Islet amyloid polypeptide-induced membrane leakage involves uptake of lipids by forming amyloid fibers. *FEBS Lett.*, 577: 117–120.

[15] Chaudhary, N., Nagaraj, R. (2009). Hen lysozyme amyloid fibrils induce aggregation of erythrocytes and lipid vesicles. *Mol. Cell. Biochem.*, 328: 209–215.

[16] Xue, W.-F., Hellewell, A. L., Gosal, W. S., Homans, S. W., Hewitt, E. W., Radford, S. E. (2009). Fibril fragmentation enhances amyloid cytotoxicity. *J. Biol. Chem.*, 284: 34272–34282.

[17] Ma, J. W., Zhao, L., Zhao, D. S., Liu, Q., Liu, C., Wu, W. H., Chen, Y. X., Zhao, Y. F., Li, Y. M. (2012). A covalently reactive group-modified peptide that specifically reacts with lysine16 in amyloid β. *Chem. Commun.*, 48: 10565–10567.

[18] Stefani, M., Rigacci, S. (2014). Beneficial properties of natural phenols: highlight on protection against pathological conditions associated with amyloid aggregation. *Biofactors*, 40: 482–493.

[19] Morshedi, D., Rezaei-Ghaleh, N., Ebrahim-Habibi, A., Ahmadian, S., Nemat-Gorgani, M. (2007). Inhibition of amyloid fibrillation of lysozyme by indole derivatives – possible mechanism of action. *FEBS J.*, 274: 6415–6425.

[20] Mao, S. S., DiMuzio, J., McHale, C., Burlein, C., Olsen, D., Carroll, S. S. (2008). A time-resolved, internally quenched fluorescence assay to characterize inhibition of hepatitis C virus nonstructural protein 3–4A protease at low enzyme concentrations. *Anal. Biochem.*, 373: 1–8.

[21] Rajasekhar, K., Chakrabarti, M., Govindaraju, T. (2015). Function and toxicity of amyloid beta and recent therapeutic interventions targeting amyloid beta in Alzheimer's disease. *Chem. Commun.*, 51: 13434–13450.

[22] Dutta, M., Kumar, M. V. S. (2015). Inhibition of Aβ aggregation in Alzheimer's disease using the poly-ion short single stranded DNA: in silico study. *J. Biomol. Struct. and Dynamics*, 33: Issue sup 1.

[23] Mishra, R., Bulic, sB., Sellin, D., Jha, S., Waldmann, H., Winter, R. (2008). Small-molecule inhibitors of islet amyloid polypeptide fibril formation. *Angew. Chem. Int. Ed. Engl.*, 47: 4679–4782.

[24] Yang, F., Lim, G. P., Begum, A. N., Ubeda, O. J., Simmons, M. R., Ambegaokar, S. S., Chen, P. P., Kayed, R., Glabe, C. G., Frautschy, S. A., Cole, G. M. (2004). Curcumin inhibits formation of amyloid beta oligomers and fibrils, binds plaques, and reduces amyloid in vivo. *J. Biol. Chem.*, 280: 5892–58901.

[25] Sun, A. Y., Wang, Q., Simonyi, A., Sun, G. Y. (2010). Resveratrol as a therapeutic agent for neurodegenerative diseases. *Mol. Neurobiol.*, 41: 375–383.

[26] Chang, E., Congdon, E. E., Honson, N. S., Duff, K. E., Kuret, J. (2009). Structure-activity relationship of cyanine tau aggregation inhibitors. *J. Med. Chem.*, 52: 3539–3547.

[27] Volkova, K. D., Kovalska, V. B., Inshin, D., Slominskii, Y. L., Tolmachev, O. I., Yarmoluk, S. M. (2011). Novel fluorescent trimethine cyanine dye 7519 for amyloid fibril inhibition assay. *Biotech. Histochem.*, 86: 188–191.

[28] Feng, B. Y., Simeonov, A., Jadhav, A., Babaoglu, K., Inglese, J., Shoichet, B. K., Austin, C. P. (2007). A high-throughput screen for aggregation-based inhibition in a large compound library. *J. Med. Chem.*, 50: 2385–2390.

[29] Shikama, Y., Kitazawa, J., Yagihashi, N., Uehara, O., Murata, Y., Yajima, N., Wada, R., Yagihashi, S. (2010). Localized amyloidosis at the site of repeated insulin injection in a diabetic patient. *Intern. Med.*, 49: 397–401.

[30] Brange, J., Andersen, L., Laursen, E. D., Meyn, G., Rasmussen, E. (1997). Toward understanding insulin fibrillation. *J. Pharm. Sci.*, 86: 517–25.

[31] Gupta, Y., Singla, G., Singla, R. (2015). Insulin-derived amyloidosis. *Indian J. Endocrinol. Metab.*, 19: 174–177.

[32] Gadjev, N. I., Deligeorgiev, T. G., Kim, S. H. (1999). Preparation of monomethine cyanine dyes as noncovalent labels for nucleic acids. *Dyes and Pigm.*, 40: 181–186.

[33] Fürstenberg, A., Julliard, M. D., Deligeorgiev, T. G., Gadjev, N. I., Vasilev, A. A., Vauthey, E. (2005). Ultrafast excited-state dynamics of DNA fluorescent intercalators: new insight into the fluorescence enhancement mechanism. *J. Am. Chem. Soc.*, 128: 7661–7669.

[34] Kurutos, A., Crnolatac, I., Orehovec, I., Gadjev, N., Piantanida, I., Deligeorgiev, T. (2016). Novel synthetic approach to asymmetric monocationic trimethine cyanine dyes derived from N-ethyl quinolinum moiety. Combined fluorescent and ICD probes for AT-DNA labeling. *J. Luminesc.*, 174: 70–76.

[35] Kurutos, A., Ryzhova, O., Trusova, V., Gorbenko, G., Gadjev, N., Deligeorgiev, T. (2016). Symmetric meso-chloro-substituted pentamethine cyanine dyes containing benzothiazolyl/ benzoselenazolyl chromophores novel synthetic approach and studies on photophysical properties upon interaction with bio-objects. *J. Fluoresc.*, 26: 177–187.

[36] Kurutos, A., Ryzhova, O., Tarabara, U., Trusova, V., Gorbenko, G., Gadjev, N., Deligeorgiev, T. (2016). Novel synthetic approach to near-infrared heptamethine cyanine dyes and spectroscopic characterization in presence of biological molecules. *J. Photochem. Photobiol. A: Chem.*, 328: 87–96.

[37] Deligeorgiev, T., Lesev, N., Kaloyanova, S. (2011) An improved method for tricyanovinilation of aromatic amines under ultrasound irradiation. *Dyes and Pigm.*, 91:74–78.

[38] Muzaffar, M., Ahmad, A. (2011). The mechanism of enhanced insulin amyloid fibril formation by NaCl is better explained by a conformational change model. *PLoS One*, 6: e27906.

[39] Adachi, E., Nakajima, H., Mizuguchi, C., Dhanasekaran, P., Kawashima, H., Nagao, K., Akaji, K., Lund-Katz, S., Phillips, M. C., Saito, H. (2013). Dual role of an N-terminal amyloidogenic mutation in apolipoprotein A-I. *J. Biol. Chem.*, 288: 2848–2856.

[40] Vus, K., Girych, M., Trusova, V., Gorbenko, G., Kurutos, A., Vasilev, A., Gadjev, N., Deligeorgiev, T. (2019). Cyanine dyes derived inhibition of insulin fibrillization. *J. Mol. Liq.*, 276: 541–552.

[41] Schmidt, M. W., Baldridge, K. K., Boatz, J. A., Elber, S. T., Gordon, M. S., Jensen, J. H., Koseki, S., Matsunaga, N., Nguyen, K. A., Su, S., Windus, T. L., Dupuis, M., Montgomery, J. A. (1993). General atomic and molecular electronic structure system. *J. Comput. Chem.*, 14: 1347–1363.

[42] Ertl, P., Rohde, B., Selzer, P. (2000). Fast calculation of molecular polar surface area as a sum of fragment based contributions and its application to the prediction of drug transport properties. *J. Med. Chem.*, 43: 3714–3717.

[43] Tetko, I. V., Tanchuk, V. Y. (2002). Application of associative neural networks for prediction of lipophilicity in ALOGPS 2.1 program. *J. Chem. Inf. Comput. Sci.*, 42: 1136–1145.

[44] Ivanova, M. I., Sievers, S. A., Sawaya, M. R., Wall, J. S., Eisenberg, D. (2009). Molecular basis for insulin fibril assembly. *Proc. Natl. Acad. Sci. U. S. A.*, 106: 18990–19995.

[45] Dodson, G., Steiner, D. (1998). The role of assembly in insulin's biosynthesis. *Curr. Opin. Struck. Biol.*, 8: 189–194.

[46] Patonay, G., Kim, J. S., Kodagahally, R., Strekowski, L. (2005). Spectroscopic study of a novel bis(heptamethine cyanine) dye and its

interaction with human serum albumin. *Appl. Spectrosc.*, 59: 682–690.

[47] Pisoni, D. S., Todeschini, L., Borges, A. C., Petzhold, C. L., Rodembusch, F. S., Campo, L. F. (2014). Symmetrical and asymmetrical cyanine dyes. Synthesis, spectral properties, and BSA association study. *J. Org. Chem.*, 79: 5511–5520.

[48] Rožman, A., Crnolatac, I., Deligeorgiev, T., Piantanida, I. (2019). Strong impact of chloro substituent on TOTO and YOYO ds-DNA/RNA sensing. *J. Luminesc.*, 205: 87–96.

[49] Vus, K., Tarabara, U., Kurutos, A., Ryzhova, O., Gorbenko, G., Trusova, V., Gadjev, N., Deligeorgiev, T. (2017). Aggregation behavior of novel heptamethine cyanine dyes upon their binding to native and fibrillar lysozyme. *Mol. Biosyst.*, 13: 970–980.

[50] Zhou, L. C., Zhao, G. J., Liu, J. F., Han, K. L., Wu, Y. K., Peng, X. J., Sunc, M.-T. (2007). The charge transfer mechanism and spectral properties of a near-infrared heptamethine cyanine dye in alcoholic and aprotic solvents. *J. Photochem. Photobiol. A: Chem.*, 187: 305–310.

[51] Vanquelef, E., Simon, S., Marquant, G., Garcia, E., Klimerak, G., Delepine, J. C., Cieplak, P., Dupradeau, F. Y. (2011). R.E.D. Server: a web service for deriving RESP and ESP charges and building force field libraries for new molecules and molecular fragments. *Nucleic Acids Res.*, 39: W511–517.

[52] Merlini, G., Bellotti, V. (2003). Molecular mechanisms of amyloidosis. *N. Engl. J. Med.*, 349: 583–596.

[53] Mahalka, A. K., Maury, C. P. J., Kinnunen, P. K. J. (2011). 1-Palmitoyl-2-(9′-oxononanoyl)-sn-glycero-3-phosphocholine, an oxidized phospholipid, accelerates finnish type familial gelsolin amyloidosis in vitro. *Biochemistry*, 50: 4877–4889.

[54] Kotormán, M., Kelemen, Z., Kasi, P. B., Nemcsók, J. (2018). Inhibition of the formation of amyloid-like fibrils using herbal extracts. *Acta Biol. Hung.*, 69: 125–134.

[55] Nielsen, L., Khurana, R., Coats, A., Frokjaer, S., Brange, J., Vyas, S., Uversky, V. N., Fink, A. L. (2001). Effect of environmental

46 *Kateryna Vus, Uliana Tarabara, Olga Zhytniakivska et al.*

factors on the kinetics of insulin fibril formation: elucidation of the molecular mechanism. *Biochemistry*, 40: 6036–6046.

[56] Cohen, F. E., Kelly, J. W. (2003). Therapeutic approaches to protein-misfolding diseases. *Nature*, 426: 905–909.

[57] Miroy, G. J., Lai, Z., Lashuel, H. A., Peterson, S. A., Strang, C., Kelly, J. W. (1996). Inhibiting transthyretin amyloid fibril formation via protein stabilization. *P. N. A. S.*, 93: 15051–15056.

[58] Biancalana, M., Koide, S. (2010). Molecular mechanism of Thioflavin-T binding to amyloid fibrils. *Biochim. Biophys. Acta.*, 1804: 1405–1412.

[59] Maskevich, A. A, Stsiapura, V. I., Kuzmitsky, V. A., Kuznetsova, I. M., Povarova, O. I., Uversky, V. N., Turoverov, K. K. (2007). Spectral properties of thioflavin T in solvents with different dielectric properties and in a fibril-incorporated form. *J. Proteome Res.*, 6: 1392–401.

[60] Pithadia, A., Brender, J. R., Fierke, C. A., Ramamoorthy, A. (2016). Inhibition of IAPP aggregation and toxicity by natural products and derivatives. *J. Diabetes. Res.*, 2046327.

[61] Volkova, K. D., Kovalska, V. B., Inshin, D., Slominskii, Y. L., Tolmachev, O. I., Yarmoluk, S. M. (2011). Novel fluorescent trimethine cyanine dye 7519 for amyloid fibril inhibition assay. *Biotech. Histochem.*, 86: 188–191.

[62] Volkova, K. D., Kovalska, V. B., Balanda, A. O., Losytskyy, M. Y., Golub, A. G., Vermeij, R. J., Subramaniam, V., Tolmachev, O. I., Yarmoluk, S. M. (2008). Specific fluorescent detection of fibrillar alpha-synuclein using mono- and trimethine cyanine dyes. *Bioorg. Med. Chem.*, 16: 1452–1459.

[63] Shinde, M. N., Barooah, N., Bhasikuttan, A. C., Mohanty, J. (2016). Inhibition and disintegration of insulin amyloid fibrils: a facile supramolecular strategy with p-sulfonatocalixarenes. *Chem. Commun.*, 52: 2992–2995.

[64] Groenning, M. (2010). Binding mode of thioflavin T and other molecular probes in the context of amyloid fibrils-current status. *J. Chem. Biol.*, 3: 1–18.

Novel Cyanine Dyes as Inhibitors of Insulin Fibrillization 47

[65] Kuznetsova, I. M., Sulatskaya, A. I., Uversky, V. N., Turoverov, K. K. (2012). Analyzing thioflavin T binding to amyloid fibrils by an equilibrium microdialysis-based technique. *PLoS One*, 7: e30724.

[66] Mishra, R., Sjölander, D., Hammarström, P. (2011). Spectroscopic characterization of diverse amyloid fibrils in vitro by the fluorescent dye Nile red. *Mol. Biosyst.*, 7: 1232–1240.

[67] Mauro, M., Craparo, E. F., Podesta, A., Bulone, D., Carotta, R., Martorana, V., Tianaa, G., BiagioSan, P. L. (2007). Kinetics of different processes in human insulin amyloid formation. *J. Mol. Biol.*, 366: 258–274.Iannuzzi, C., Borriello, M., Portaccio, M., Irace, G., Sirangelo, I. (2017). Insights into insulin fibril assembly at physiological and acidic pH and related amyloid intrinsic fluorescence. *Int. J. Mol. Sci.*, 18: E2551.

[68] Jansen, R., Dzwolak, W., Winter, R. (2005). Amyloidogenic self-assembly of insulin aggregates probed by high resolution atomic force microscopy. *Biophys. J.*, 88: 1344–1353.

[69] Dubey, K., Anand, B. G., Shekhawat, D. S., Kar, K. (2017). Eugenol prevents amyloid formation of proteins and inhibits amyloid-induced hemolysis. *Sci. Rep.*, 7: 40744.

[70] Uversky, V. N. (2010). Mysterious oligomerization of the amyloidogenic proteins. *FEBS J.*, 277: 2940–2953.

[71] Krebs, M. R., Bromley, E. H., Rogers, S. S., Donald, A. M. (2005). The mechanism of amyloid spherulite formation by bovine insulin. *Biophys. J.*, 88: 2013–2021.

[72] Chirita, C., Necula, M., Kuret, J. (2004). Ligand-dependent inhibition and reversal of tau filament formation. *Biochemistry*, 43: 2879–2887.

[73] Masuda, M., Suzuki, N., Taniguchi, S., Oikawa, T., Nonaka, T., Iwatsubo, T., Hisanaga, S., Goedert, M., Hasegawa, M. (2006). Small molecule inhibitors of alpha-synuclein filament assembly. *Biochemistry*, 45: 6085–6094.

[74] Hou, T. J., Xu, X. J. (2003). ADME evaluation in drug discovery. 3. Modeling blood-brain barrier partitioning using simple molecular descriptors. *J. Chem. Inf. Comput. Sci.*, 43: 2137–2152.

[75] Ert, P., Rohde, B., Selzer, P. (2000). Fast calculation of molecular polar surface area as a sum of fragment-based contributions and its application to the prediction of drug transport properties. *J. Med. Chem.*, 43: 3714–3717.

[76] Schafer, K. N., Cisek, K., Huseby, C. J., Chang, E., Kuret, J. (2013). Structural determinants of Tau aggregation inhibitor potency. *J. Biol. Chem.*, 288: 32599–32611.

[77] Mahmoodabadi, N., Ajloo, D. (2016). QSAR, docking, and Molecular dynamic studies on the polyphenolic as inhibitors of β-amyloid aggregation. *Med. Chem. Res.*, 25: 2104–2118.

[78] Hannah, K. C., Armitage, B. A. (2004). DNA-templated assembly of helical cyanine dye aggregates: a supramolecular chain polymerization. *Acc. Chem. Res.*, 37: 845–853.

[79] Wu, C., Wang, Z., Lei, H., Zhang, W., Duan, Y. (2007). Dual binding modes of Congo red to amyloid protofibril surface observed in molecular dynamics simulations. *J. Am. Chem. Soc.*, 129: 1225–1232.

[80] Neidari, N., Saboury, A. A., Meratan, A. A., Karami, L., Sawyer, L., Kaboudin, B., Jooyan, N., Ghasemi, A. (2018). Polyphenolic self-association accounts for redirecting a high-yielding amyloid aggregation. *J. Mol. Liq.*, 266: 291–298.

[81] Karelson, M., Lobanov, V. S., Katritzky, A. R. (1996). Quantum-chemical descriptors in QSAR/QSPR studies. *Chem. Rev.*, 96: 1027–1044.

[82] Parthasarathy, V., Pandey, R., Stolte, M., Ghosh, S., Castet, F., Würthner, F., Das, P. K., Blanchard-Desce, M. (2015). Combination of cyanine behaviour and giant hyperpolarisability in novel merocyanine dyes: Beyond the bond length alternation (BLA) paradigm. *Chemistry*, 21: 14211–14217.

[83] Puyad, A. L., Chaitanya, G. K., Thomas, A., Paramasivam, M., Bhanuprakash, K. (2013). DFT studies of squarylium and core-substituted squarylium dye derivatives: understanding the causes of the additional shorter wavelength absorption in the latter. *J. Physical. Organic. Chem.*, 26: 37–46.

[84] Steinmann, D., Nauser, T., Koppenol, W. H. (2010). Selenium and sulfur in exchange reactions: a comparative study. *J. Org. Chem.*, 75: 6696–6699.

[85] Manjare, S. T., Kim, Y., Churchill, D. G. (2014). Selenium- and tellurium-containing fluorescent molecular probes for the detection of biologically important analytes. *Acc. Chem. Res.*, 47: 2985–2998.

[86] Yu, F., Li, P., Li, G., Zhao, G., Chu, T., Han, K. (2011). A near-IR reversible fluorescent probe modulated by selenium for monitoring peroxynitrite and imaging in living cells. *J. Am. Chem. Soc.*, 133: 11030–11033.

[87] Wainwright, M. (2004). Photodynamic therapy – from dyestuffs to high-tech clinical practice. *Coloration Technol.*, 34: 95–109.

[88] Sakagashira, S., Hiddinga, H. J., Tateishi, K., Sanke, T., Hanabusa, T., Nanjo, K., Eberhardt, N. L. (2000). S20G mutant amylin exhibits increased in vitro amyloidogenicity and increased intracellular cytotoxicity compared to wild-type amylin. *Am. J. Pathol.*, 157: 2101–2109.

[89] Cattaneo, M., Lecchi, A., Randi, A. M., McGregor, J. L., Mannucci, P. M. (1992). Identification of a new congenital defect of platelet function characterized by severe impairment of platelet responses to adenosine diphosphate. *Blood*, 80: 2787–2796.

[90] Sawaya, M. R., Sambashivan, S., Nelson, R., Ivanova, M. I., Sievers, S. A., Apostol, M. I., Thompson, M. J., Balbirnie, M., Wiltzius, J. J., McFarlane, H. T., Madsen, A., Riekel, C., Eisenberg, D. (2007). Atomic structures of amyloid cross-beta spines reveal varied steric zippers. *Nature*, 447: 453–457.

[91] Jiang, P., Li, W., Shea, J.-E., Mu, Y. (2011). Resveratrol inhibits the formation of multiple-layered β-sheet oligomers of the human islet amyloid polypeptide segment 22–27. *Biophys. J.*, 100: 1550–1558.

[92] Hong, Y., Meng, L., Chen, S., Leung, C. W., Da, L. T., Faisal, M., Silva, D. A., Liu, J., Lam, J. W., Huang, X., Tang, B. Z. (2012). Monitoring and inhibition of insulin fibrillation by a small organic fluorogen with aggregation-induced emission characteristics. *J. Am. Chem. Soc.*, 134: 1680–1689.

[93] Wu, C., Lei, H., Wang, Z., Zhang, W., Duan, Y. (2006). Phenol red interacts with the protofibril-like oligomers of an amyloidogenic hexapeptide NFGAIL through both hydrophobic and aromatic contacts. *Biophys. J.*, 91: 3664–3672.

[94] Feng, B. Y., Simeonov, A., Jadhav, A., Babaoglu, K., Inglese, J., Shoichet, B. K., Austin, C. P. (2007). A high-throughput screen for aggregation-based inhibition in a large compound library. *J. Med. Chem.*, 50: 2385–2390.

[95] Basu, A., Kumar, S. G. (2017). Binding and inhibitory effect of the dyes Amaranth and Tartrazine on amyloid fibrillation in lysozyme. *J. Phys. Chem. B*, 121: 1222–1239.

[96] Arora, A., Ha, C., Park, C. B. (2004). Inhibition of insulin amyloid formation by small stress molecules. *FEBS Lett.*, 564: 121–125.

[97] Nielsen, L., Frokjaer, S., Brange, J., Uversky, V. N., Fink, A. L. (2001). Probing the mechanism of insulin fibril formation with insulin mutants. *Biochemistry*, 40: 8397–8409.

[98] Kuret, J., Chirita, C. N., Congdon, E. E., Kannanayakal, T., Li, G., Necula, M., Yin, H., Zhong, Q. (2005). Pathways of tau fibrillization. *Biochim. Biophys. Acta*, 1739: 167–178.

[99] Kovalska, V., Losytskyy, M., Chernii, V., Volkova, K., Tretyakova, I., Cherepanov, V., Yarmoluk, S., Volkov, S. (2012). Studies of anti-fibrillogenic activity of phthalocyanines of zirconium containing out-of-plane ligands. *Bioorg. Med. Chem.*, 20: 330–334.

[100] Huang, K., Maiti, N. C., Phillips, N. B., Carey, P. R., Weiss, M. A. (2006). Structure-specific effects of protein topology on cross-beta assembly: studies of insulin fibrillation. *Biochemistry*, 45: 10278–10293.

[101] Hua, Q. X., Weiss, M. A. (2004). Mechanism of insulin fibrillation: the structure of insulin under amyloidogenic conditions resembles a protein-folding intermediate. *J. Biol. Chem.*, 279: 21449–21460.

[102] Moraitakis, G., Goodfellow, J. M. (2003). Simulations of human lysozyme: probing the conformations triggering amyloidosis. *Biophys. J.*, 84: 2149–2158.

[103] Muhammad, E. F., Adan, R., Latif, M. A. M., Rahman, M. B. A. (2016). Theoretical investigation on insulin dimer-β-cyclodextrin interactions using docking and molecular dynamics simulation. *J. Incl. Phenom. Macrocycl. Chem.*, 84: 1–10.

[104] Porat, Y., Abramowitz, A., Gazit, E. (2006). Inhibition of amyloid fibril formation by polyphenols: structural similarity and aromatic interactions as a common inhibition mechanism. *Chem. Biol. Drug. Des.*, 67: 27–37.

In: Cyanine Dyes
Editor: Douglas Zimmerman

ISBN: 978-1-53616-239-4
© 2019 Nova Science Publishers, Inc.

Chapter 2

INTERACTIONS BETWEEN THE NOVEL CYANINE DYES AND BIOLOGICAL MACROMOLECULES

Olga Zhytniakivska[1,], Kateryna Vus[1], Valeriya Trusova[1], Uliana Tarabara[1], Galyna Gorbenko[1], Atanas Kurutos[2], Nikolai Gadjev[3] and Todor Deligeorgiev[3]*

[1]Department of Nuclear and Medical Physics,
V. N. Karazin Kharkiv National University, Kharkiv, Ukraine
[2]Institute of Organic Chemistry with Centre of Phytochemistry,
Bulgarian Academy of Sciences, Sofia, Bulgaria
[3]Department of Pharmaceutical and Applied Organic Chemistry,
Sofia University "St. Kliment Ohridski," Sofia, Bulgaria

ABSTRACT

Cyanine dyes, photosensitive nitrogen-containing heterocyclic structures, are of special interest in the chemistry of dyes and pigments due to their extensive use in various fields of science, technology,

---

* Corresponding Author's E-mail: olya_zhitniakivska@yahoo.com.

pharmacology and biomedicine. Despite considerable progress achieved in the field of cyanine synthesis and application, the development and characterization of new fluorophores of cyanine family is of great importance. In the present study, we described the spectral properties of a series of monomethine, penthamethine and heptamethine cyanine dyes with an accent on their applicability to non-covalent labeling of biological macromolecules. The spectral characteristics of the novel dyes were studied in the aqueous media, in the presence of nucleic acids, proteins and model lipid membranes. The cyanine dyes under study, except of monomethines, undergo H-type self-association when free in buffer solution. Based on the results of absorption measurements, it was concluded that the examined cyanines are capable of associating with nucleic acids, proteins and lipid bilayers and this process is accompanied by the changes in the dye aggregation pattern. A tentative model for the heptamethine aggregation behavior in the presence of fibrillar and native lysozyme was formulated, assuming the dye monomer and dimer association with the protein, in addition to the dye aggregation in buffer. The association constants and stoichiometry of the dye-fibril complexation have been evaluated. The hypothesis describing the protein-dye binding mode has been proposed. The fluorescence studies showed that most dyes have negligible emission in the buffer solution as well as in the presence of DNA, proteins and model membranes, presumably due to a non-emissive nature of the dye aggregates. The exception are the monomethines exhibiting a significant emission increase upon the binding to the double-stranded DNA (dsDNA). The binding parameters for the monomethine complexation with dsDNA have been determined which are consistent with the intercalative DNA-dye binding mode. The recovered pronounced changes in the spectral responses of novel cyanines to DNA, native or fibrillar proteins and lipid bilayers allowed us to recommend these dyes for the detection and characterization of biomolecules as complementary to the existing biomarkers.

Keywords: cyanine dyes, aggregation, DNA, liposomes, amyloid fibrils

INTRODUCTION

Cyanine dyes, photosensitive nitrogen-containing heterocyclic structures, are of special interest in the chemistry of dyes and pigments due to their extensive use in various fields of science, technology, pharmacology and biomedicine. More specifically, cyanines have found numerous applications as spectral sensitizers for silver halide emulsion in photographic industry for coloured and non-coloured films [1, 2], as mode-locking compounds in laser technologies [3-5] and dye-sensitized solar cells [6, 7], as photopolymerization initiators [8, 9] and corrosion inhibitors [10], to name only a few. The above applications of cyanine dyes are dictated by their favorable spectral properties, *viz*: i) intensive absorption in the spectral region from UV to NIR; ii) high sensitivity to the properties of their microenvironment; iii) ability to undergo photoinduced geometrical isomerization in the excited state and to change the absorption and fluorescence properties; iv) capability to convert light energy to electricity, etc. Likewise, these compounds have been broadly employed as anti-tumor agents in medicine [11, 12], bactericidal agents in pharmacy [13] and as optical probes for membrane potential, lipid bilayer structure and dynamics [14-16], as strong oxidizing agents [17] and probes for the covalent labeling of biopolymers [18-20]. However, the recent trends towards application of cyanine dyes as non-covalent labels for the detection of proteins [21-23], nucleic acids [24-27] and lipids [28, 29] are observed. The non-covalent labelling of biomolecules is dictated mainly by the huge enhancement in emission upon binding to biopolymers as a result of the formation of stable dye-biopolymer complexes by means of the elegant balance between the hydrophobic and electrostatic interactions [25, 26, 30, 31].

It is worth nothing that the applications of cyanines are not limited by the electron properties of a single dye molecule but are also based on the specific characteristics of their assemblies. It is generally known that cyanine dyes possess very high polarizability of the π-electron system along the polymethine group in the ground state, which gives rise to strong dispersion forces between two cyanine molecules in solution [32]. These dispersion forces are believed to control the formation of extended aggregates of cyanines [33, 34]. Depending on the angle of molecular slippage, α, the self-association of cyanine dyes results in the formation of molecular aggregates of different structures, known as H-and J-aggregates [35, 36]. In particular, when $\alpha=90°$ the molecules are in a parallel orientation and form "card-pack" arrangement. Meanwhile, when $\alpha=0°$ a linear molecular orientation is observed, namely "end-to-end" stacking [35]. In a general case, a large molecular slippage ($\alpha<32°$) leads to the self-organization of "brickwork" arrangements (J-aggregates), while a small one ($\alpha>32°$) is characterized by "card-pack" structures (H-aggregates). The impact of the molecular geometry of the aggregate on the optical response of cyanine dyes was described early by Kasha [37]. According to the exciton model for molecular aggregates, the interactions between the neighboring transition dipoles of tightly packed molecules generate a splitting of the excited state into excitonic levels that are shared between all molecules within arrangement [37]. When two dipoles are in a side-to-side orientation (H-aggregates), the energy of the allowed state corresponding to in phase transition dipoles is increased by the repulsive electrostatic interactions between the transition dipoles resulting in the shift of absorption spectra to short-wavelengths region relative to the monomer band [38]. In contrast, for the aggregates with "head-to-tail" orientations (J-aggregates), the sign of the resonant excitonic coupling is negative, resulting in a red shift in the absorption spectrum [39].

A good wealth of reports has been devoted to the analysis of the process of cyanine aggregation. The structure of of the dye was shown to play a pivotal role in controlling the aggregation propensity of the fluorophore. Specifically, the substituents in heterocyclic residues and the N-alkyl chain length have been found to determine the type of aggregates formed [34, 40, 41]. Importantly, the changes in the environmental conditions may shift the monomer–aggregate equilibria, resulting in the mutual conversions between the dye species. Accordingly, the dye concentration [42], solvent polarity [43], temperature [44], presence of polyelectrolytes [45, 46], surfactants [47], are among the factors contributing to the formation of highly organized molecular arrangements. Another factor that was shown to affect the formation of cyanine aggregates is high ionic strength [33]. It has been found that at high salt concentration the aggregates are more energetically stable than the monomeric dyes due to the formation of ion pairs between the cationic dyes and counterions [33]. More interesting, biomolecules, such as proteins, lipids and nucleic acids, have a significant impact on the aggregation properties of the organic dye molecules [42, 48-51]. Specifically, DNA was reported to serve as a template for the growth of helical arrays of cyanine dyes which occurs by a cooperative, chain-growth mechanism [42, 48, 52]. The role of proteins as a scaffold for the cyanine aggregation or disaggregation was also described in a number of works [50, 51]. Particularly, the interaction of bis-heptamethine cyanine dye with the human serum albumin was accompanied by the destruction of the dye aggregates manifesting itself in the fluorescence intensity increase [53]. It was also shown, that the ability of phtalocyanines to inhibit the amyloid fibril formation is connected with their tendency to form the aggregates in solution due to the strong stacking interactions with aromatic amino acids [54].

During the past decade our research efforts have been focused on the synthesis and characterization of the possible biological application of novel derivatives of cyanine family [25, 26, 51, 55]. In the present study, we described the spectral properties of a series of monomethine, penthamethine and heptamethine cyanine dyes with an accent on their applicability to the non-covalent labeling of biological macromolecules. More specifically, our goals were: i) to characterize the photophysical properties of cyanine dyes in buffer solution and in the presence of biopolymers; ii) to examine the impact of DNA, proteins and lipids on the aggregation properties of the examined dyes; iii) to identify the possible binding sites of cyanines.

MATERIALS AND METHODS

Experimental

Hen egg white lysozyme, bovine serum albumin (BSA), calf thymus DNA and Tris-HCl were obtained from Sigma (USA). Bovine heart cardiolipin, 1-palmitoyl-2-oleoyl-*sn*-glycero-3-phosphocholine (PC) and 1-palmitoyl-2-oleoyl-*sn*-glycero-3-phospho-*rac*-glycerol (PG) were from

Avanti Polar Lipids (Alabaster, AL). Cyanine probes (Figure 1, Table 1) were synthesized as described previously [56, 57]. All other starting materials and solvents were commercial products of analytical grade and were used without further purification.

Preparation of Working Solutions

Stock solutions of cyanines were prepared by dissolving the dye in methanol or DMSO, then diluted in 10 mM Tris-HCI buffer (pH 7.4) and used for spectroscopic measurements. The concentration of dyes was determined spectrophotometrically, using their extinction coefficients (Table 1). The working solutions of the DNA, BSA, native and fibrillar lysozyme were prepared in 10 mM Tris-HCl buffer, pH 7.4. The concentrations of DNA and BSA were determined spectrophotometrically using their molar absorptivities $\varepsilon_{260} = 6.4 \times 10^3 \, M^{-1} cm^{-1}$, and $\varepsilon_{276} = 4.25 \times 10^4 \, M^{-1} cm^{-1}$ for the DNA and BSA, respectively.

Lysozyme amyloid fibrils were grown by incubation of the protein solution (10 mg/ml) in 10 mM glycine buffer (pH 2) at 60 °C during 14 days [58]. Lipid vesicles composed of PC and PC mixtures with CL and PG were prepared using the extrusion technique. A thin lipid film was first formed of the lipid mixtures in chloroform by removing the solvent under a stream of nitrogen. The dry lipid residues were subsequently hydrated with 20 mM HEPES, 0.1 mM EDTA, pH 7.4 at room temperature to yield lipid concentration of 1 mM. Thereafter, lipid suspension was extruded through a 100 nm pore size polycarbonate filter (Millipore, Bedford, USA). In this way, 3 types of lipid vesicles containing PC and 67 mol% CL and mol% 20 PG, with the content of phosphate being identical for all liposome preparations. Hereafter, liposomes containing 20 mol% PG are referred to as PG, while liposomes bearing 67 mol% CL are denoted as CL, respectively.

Spectroscopic Measurements

To record the absorption of cyanine dyes in the presence of biomacromolecules and model membranes, appropriate amounts of the stock solution of DNA, BSA, native or fibrillar lysozyme and liposomes were added to each dye in buffer. The absorption spectra were recorded using the spectrophotometer CM-2203 (SOLAR, Belarus) in quartz cells of 1.0 cm path length. The fluorescence measurements were performed at $25 \degree C$ using the spectrofluorimeter Perkin Elmer LS45. The fluorescence spectra of cyanine dyes were recorded with the excitation taken at the maximum of absorption spectra (Table 1). To qualitatively analyze the dye-DNA and dye-lysozyme binding, dye solutions were titrated with the double stranded DNA and protein, respectively.

Binding Model

The thermodynamic analysis of the cyanine-DNA interactions was performed in terms of the McGhee & von Hippel excluded site model allowing the calculation of the binding constant and the stoichiometry [59]:

$$\frac{B}{F} = K_a P \left(1 - \frac{nB}{P}\right) \left[\frac{1 - (nB/P)}{1 - (n-1)(nB/P)}\right]^{n-1} \tag{1}$$

where B and F are the concentrations of the bound and free dye, respectively, P is the DNA phosphate concentration, K_a denominates the association constant and n represents the site exclusion parameter (i.e., the number of base pairs excluded by the binding of a single ligand molecule). The values of K_a and n were estimated using the nonlinear least-square fitting procedure.

Docking Studies

Crystal structures of hen egg white lysozyme (PDB ID: 3A8Z) and bovine serum albumin (PDB ID: 4F5S) were taken from the Protein Data Bank. Lysozyme fibril was built from the K-peptide, GILQINSRW (residues 54–62 of the wild-type protein), using CreateFibril tool as described previously [60]. The geometry of cyanine dyes was optimized in Avogadro. The docking models of the dye dimers and the complexes between the dye monomer (dimer) and native bovine serum albumin or lysozyme, or fibrillar lysozyme were obtained using the PatchDock algorithm that is a user-friendly tool for calculation of the optimal structures of the protein-drug and protein-protein complexes [61]. The online-available program searches the transformations of the two interacting molecules (assuming the proteins to be rigid bodies), revealing the maximized surface shape complementarity and minimized number of steric clashes. The top 10 obtained conformations were then refined by the FireDock algorithm, that calculates the optimal rearrangement of the side chains in the protein-ligand complex by Monte Carlo minimization of the binding score function (comprising the following important contributants – energy of ligand-protein wan der Waals interactions and desolvation free energy). The docked complexes were visualized by the Visual Molecular Dynamics (VMD) software.

RESULTS AND DISCUSSION

Spectroscopic Characterization of Cyanine Dyes in Methanol and in Buffer Solution

As seen in Figure 1 and Table 1, the cyanines **M1-M19** under investigation are assymetric monomethine dyes, containing a benzothiazole fragment. In turn, pentamethines and heptamethines are symmetric one. Cyanines **P1-P7** are benzothiazolic and benzoselenazolic heterocycles with a *meso*-chlorine substituent with respect to the polymethine chain. Symmetric heptamethine dyes **H1** and **H2** are based on integrated chloro-substituted cyclohexenyl ring. The dyes are cationic in nature due to the delocalized positive charge of the chromophore. Presented in Table 1 are the absorption maximum (λ_{max}) and molar absorptivities (ε) of the examined cyanines in methanol. Absorption spectra of monomethine dyes are typical to those observed for Thiazole Orange derivatives [62, 63] with the absorption maximum in methanol in a range from 507 to 517 nm depending on the dye structure. Absorption spectra of pentamethine and heptamethine dyes are bathochromically shifted with respect to monomethines by ~190 nm and ~340 nm, respectively, describing a well-known tendency of cyanines dyes to correspond to the lengthening of their polymethine chain by red shift of absorption band [64]. Consequently, penthamethines are characterized by broad absorption spectra within the range of 450-750 nm with the absorption maximum in a range from 645 to 657 nm, whereas absorption maxima for **H1** and **H2** were at 795 and 800 nm, respectively.

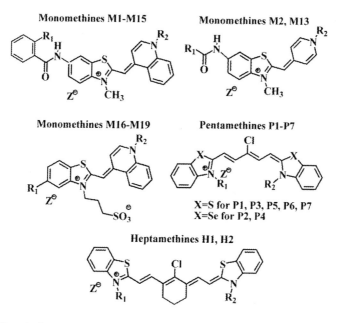

Figure 1. Chemical structures of the monomethine cyanine dyes.

Table 1. Chemical structure of dyes, their absorption maxima and extinction coefficients in methanol

Dye	R_1	R_2	Z	λ_{max} (nm)	ε (M^{-1} cm^{-1})
M1			I$^-$	519	70300
M2			Br$^-$	464	37900
M3			Br$^-$	518	81100
M4		$C_6H_5C_2H_4$	Br$^-$	519	77900
M5		$C_6H_5CH_2$	Br$^-$	519	73600
M6		C_2H_4OH	Br$^-$	516	61900
M7	Cl		Br$^-$	519	64000
M8	F		Br$^-$	519	75300
M9	F	$C_6H_5CH_2$	Br$^-$	519	93100
Dye	R_1	R_2	Z	λ_{max} (nm)	ε (M^{-1} cm^{-1})
M10	F		Br$^-$	518	68700
M11	Cl	$C_6H_5CH_2$	Br$^-$	519	85600
M12		$C_5H_{10}COOH$	Br$^-$	517	78900
M13	CH_3		I$^-$	453	66400
M14			I$^-$	521	49200
M15	F		Br$^-$	517	58600
M16			I$^-$	507	83200
M17			I$^-$	507	77200
M18	CH_3O		I$^-$	517	70700

Dye	R₁	R₂	Z	λ_{max} (nm)	ε (M⁻¹ cm⁻¹)
M19	CH₃O	(structure)	I⁻	517	59300
P1	CH₃	CH₃	I⁻	645	203550
P2	CH₃	CH₃	I⁻	655	228850
P3	C₂H₅	C₂H₅	I⁻	646	266450
P4	C₂H₅	C₂H₅	I⁻	657	253560
P5	C₆H₅CH₂	C₆H₅CH₂	I⁻	653	218860
P6	C₂H₄CN	C₂H₄CN	I⁻	652	273140
P7	C₂H₄OH	C₂H₄OH	I⁻	650	278120
H1	C₃H₅C₂H₆	C₃H₅C₂H₆	I⁻	795	288890
H2	C₆H₅CH₂	C₆H₅CH₂	I⁻	800	332163

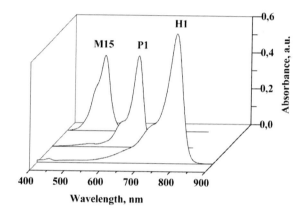

Figure 2. Typical absorption spectra of cyanine dyes in methanol. Dye concentration was 2.5 μm.

As can be seen from Figure 2, absorption spectra of all cyanine dyes under study were featured by the two-peak structure with well-defined high-intensity absorption band and a less intense sub-peak at the shortwavelength side of the band. Numerical studies indicate that the absorption spectra of cyanine dyes possess a vibronic structure with the relative intensities of 0-0 and 0-1 sub-bands being dependent on the dye structure [65]. The main absorption band of cyanines centered at λ_{max} corresponds to the 0-0 transition, whereas a second sub-band can be assigned to 0-1 transition. It should be noted that photophysical properties of cyanine dyes along with the length of polymethine chain are directly influenced by the chemical substitution of the terminal heterocyclic groups [64]. Replacement of benzoselenazolic heterocycle by benzothiazolic one caused a 11 nm bathochromic shift in the absorption maxima position for dyes **P2** and **P4** in comparison with **P1** and **P3**, respectively. Introduction of a phenyl ring, cyano- and hydroxy group into the **P5**, **P6** and **P7** structure, respectively, leads to 5-8 nm shifts in absorption maximum positions in comparison with **P1**. Likewise, the addition of one methoxy group to the benzothiazole moiety of **M18** and **M19**, resulted in 10 nm shifts in the absorption maxima position in comparison with **M16, M17**.

Table 2. Absorption maxima of pentamethine and heptamethine cyanine dyes in buffer

Dye	$^*\lambda_A^M$	λ_A^H	^{b}LogP
P1	640	583, 503	2.62
P2	654	592, 543	2.13
P3	642	585	3.37
P4	655	596	2.88
P5	653	542^a	5.81
P6	650	594	1.53
P7	647	592	1.36
H1	808	589^a	5.40
H2	815	641^a	5.51

$^*\lambda_A^H$, λ_A^M are absorption wavelength of aggregates maxima and monomeric dye form;

^{a}The most intensive maximum, in the case that this maximum does not belong to dye monomers.

^{b}LogP is lipophilicity values of the examined compounds which were calculated using the Virtual Computational chemistry laboratory (http://www.vcclab.org).

Figure 3 represents typical absorption spectra of the examined cyanine dyes in the buffer solution. In aqueous medium significant changes in the cyanine spectral behavior were observed presumably for penthamethine and heptamethine dyes whereas monomethine dyes transfer to the buffer solution was accompanied with the 2-10- nm shift of the absorption maxima without any changes in the spectrum shape indicating that monomethines **M1-M19** in the buffer solution are presented presumably in the monomeric dye form. Thiazole Orange and its derivatives have been reported to exist in the monomeric form in an aqueous media at the micromolar dye concentration [66, 67]. The changes in heptamethine and penthamethine spectra are presented in Table 2. As judged from this Table, absorption spectra of penthamethine dyes **P3**, **P4**, **P5** and **P6** were featured by the two-peak structure. The bathochromic peak (ranging from 642 nm to 655 nm) corresponds to the absorption of monomeric dye species, while the hypsochromic one (with maximum between 585 nm and 596 nm) is a hallmark of H-aggregates. Importantly, the monomer maximum prevailed over that of the aggregates.

Absorption spectra of penthamethine dyes **P1** and **P2** are characterized by three peaks, corresponding presumably to the dye monomers (640 nm and 654 nm), dimers (583 nm and 592 nm) and higher order H-aggregates (503 nm and 543 nm).

The ratio of these bands was found to depend on the dye chemical structure, but the monomer maximum also prevailed over that of the aggregate. In turn, for penthamethine dye **P5** the absorption spectrum was hypsochromically shifted with respect to those observed for other pentamethines under study indicating that H-aggregates represent the dominating fraction of the **P5** in the buffer solution. It should be noted that the similar behavior was observed for heptamethines. The highly intensive and well-defined narrow bands of **H1** and **H2** (Figure 3 panel A) were observed at 598 nm and 641 nm, respectively, whereas the monomer peaks (808 and 815 nm) were virtually unstructured and resembled the spectrum shoulder. Bis-heptamethine cyanine dyes have been reported to form H-aggregates in buffer solution [68]. Interestingly, the highest aggregation tendency was observed for dyes with relatively high Logarithmic Partition coefficients (LogP) (Table 2). Lipophilicity, an important molecular parameter, characterizes the tendency of a molecule to distribute between water and water-immiscible solvent which can be expressed by a volume or cavity term accounting for hydrophobic and dispersion forces, and polarity term determined by electrostatic interactions [69]. It is generally known that the aggregation of cyanines are controlled by the balance between the van der Waals interactions, dispersion forces of the cyanine backbone, the forces of the alkyl chain of entropic and hydrophobic nature together with H-bonding and the electrostatic interactions of the ionic groups [34]. It has been shown that the structure of the dye, particularly, the presence of substituents in heterocyclic residues and the N-alkyl chain length have a significant impact on aggregation of cyanine dyes [34, 40, 41]. Having the highest LogP values, **P5**, **H1** and **H2** with a phenyl rings in terminal group seem to be more hydrophobic with respect to other dyes under study thereby displaying more pronounced aggregation.

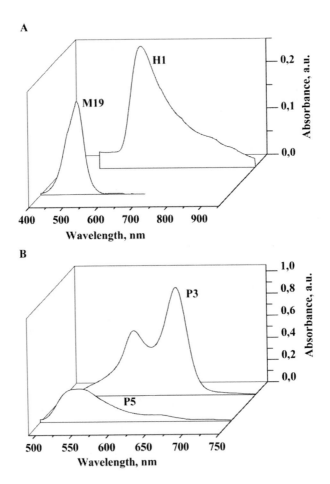

Figure 3. Typical absorption spectra of cyanines in buffer. Dyes concentration was 5 µM, 12,2 µM, 7,3 µM and 5,6 µM for M19, H1 (panel A) and P3, P5 (panel B) respectively.

The Association of Cyanine Dyes with the DNA

During the past century cyanine dyes attract ever growing interest as extrinsic fluorescent probes and labels for the quantitative sensing, rapid visualization and characterization of nucleic acids. More specifically, cyanine dyes were used for sizing and purification of the DNA fragments [70, 71], fluorescent microscopy applications [72], the DNA damage detection [73], microarray-based expression analysis [74], the DNA sequencing [75, 76], the DNA intercalation bioanalytical assays as well as for staining of the nucleic acids in electrophoresis [70, 77], to name only a few. Therefore, at the next step of our study it was logical to assess the sensitivity of cyanine dyes to the double-stranded DNA.

Interactions between Monomethine Cyanine Dyes and the Double Stranded DNA

The association of monometine dyes **M1-M19** with the double stranded DNA was accompanied with a small hypochromism along with 2-11 bathochromic shifts in the position of the absorption peak, depending on the dye structure (Table 3). The above red shift in the monomethine absorption indicates a decrese in the polarity of the **M1-M19** environment. It is well known that intercalation of small molecules into the DNA helix results in red shift of the spectra as well as hypochromism, whereas hyperchromism is a spectral feature depicting non-covalent interactions (electrostatic and groove binding) between small ligands and DNA [73, 78]. Since hypochromism was observed for **M1-M19** association with DNA, it is highly probable that these dyes intercalate into the base pair stack at the core of the double helix. Therefore, to elucidate further the interaction of monomethine cyanine dyes under study with DNA steady state fluorescence measurements were performed. The maxima of the absorption and fluorescence band for free and DNA-bound dyes, along with the absolute and relative fluorescence intensities of the free dye in TE buffer (I_0) and dye-DNA solutions (I) are given in Table 3. The dyes were found to have negligible fluorescence in TE buffer in the absence of the DNA (Table 3). It has been shown that Thiazole Orange derivatives undergo rapid deactivation from the singlet excited state by rotating around double bond joining benzothiazole and quinoline ring, a well-established decay mechanism for alkenes [79]. This energy wasting mechanism manifests itself in a dramatic reduction of fluorescence. The situation is changed when the dye is placed in the DNA solution with the restricted motional freedom that results in cyanine energy dissipation predominantly through emission *via* fluorescence. Accordingly, in the presence of dsDNA all dyes under study show enhanced fluorescence (approximately from 16 to 1033-fold increase in fluorescence intensity) thereby creating prerequisites for their use as fluorescence dyes for DNA detection and characterization. Comparison of the spectral properties of free and dsDNA-bound dyes under the same experimental conditions revealed that the most pronounced fluorescence increase is observed for dyes **M8**, **M9** and **M11**, pointing to significantly greater ability of these dyes (in comparison with the other examined dyes) to respond to the DNA-induced changes in the dye microenvironment.

Interestingly, the minor differences in cyanine structure can substantially affect the magnitude of fluorescence enhancement (I / I_0) in the presence of dsDNA. Hence, the replacement of quinoline with the pyridine one significantly supressed the fluorescence enchancement of **M2** and **M13**. Furthermore, the smallest fluorescence increase upon dye transfer from the buffer solution to the DNA was observed for monomethines containing more bulky substitutions. The above results are in an agreement with Lerman's intercalation model [80] postulating that introduction of bulky substituent into the dye molecule negatively affects fluorescence intensity of the dye-DNA complexes. Similar results were obtained by Yarmoluk and coworkers who showed that bulky substitution in the benzothiazole residue led to the significant decrease in the fluorescence enhancement [63].

In order to characterize the stability of cyanine-DNA complexes and to make an assumption about the complexation mechanisms, the DNA binding parameters of the examined monomethine cyanine dyes were obtained from the analysis of fluorescence titration data (Figure 4). Fluorescence intensity was found to increase upon the dye transfer from aqueous phase to dsDNA solution, with emission maximum being shifted towards the lower wavelengths. These findings can be explained by the reduced polarity of the dye surroundings and restricted fluorophore mobility within the DNA binding sites. To explain the results of fluorescence measurements quantitavely, the increase in fluorescence intensity in the presence and absence of DNA (ΔI) has been plotted as a function of the DNA base pairs-to-dye molar ratio (P/D) (inset of Figure 4 A), or as a function of the DNA concentration (inset of Figure 4 B). The observed binding isotherms were analyzed in terms of the McGhee & von Hippel excluded site model (Eq. (1)). The results obtained are summarized in Table 4.

Table 3. Spectral characteristics of monomethine cyanine dyes in buffer solution and in the presence of DNA obtained at a molar ratio of DNA bp/dye of 10:1

Dye	Buffer	In the presence of DNA		I_0 (a.u.)	I (a.u.)	I/I_0
	λ_{max} (nm)	λ_{max} (nm)	λ_{em} (nm)			
M1	507	519	547	2.98	832.51	279
M2	458	473	496	7.91	128.06	15
M3	517	530	547	8.35	446.34	53
M4	509	520	546	1.31	223.52	170
M5	500	520	545	1.29	348.22	270
M6	503	516	546	2.02	695.58	344
M7	504	519	543	0.89	383.41	383
M8	508	519	544	1.12	998.80	892
M9	501	519	545	0.96	992.15	1033
M10	512	525	545	2.09	710.18	340
M11	512	519	543	0.98	817.24	834
M12	509	512	543	2.45	941.71	385
M13	451	465	496	8.79	770.15	88
M14	509	512	549	0.99	183.86	186
M15	510	516	547	0.80	349.32	437
M16	507	514	532	0.9	98.4	109
M17	508	515	532	0.75	161.8	215
M18	518	515	549	1.8	299.4	158
M19	518	515	548	0.6	80.04	133

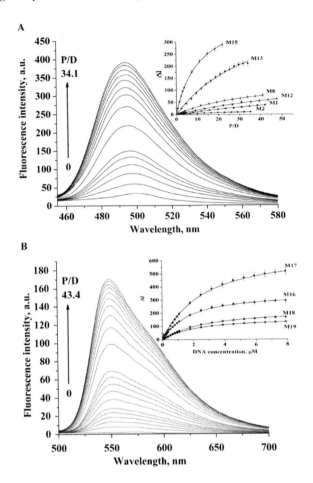

Figure 4. Representative fluorescence spectra of cyanine dyes **M13** (A) and **M18** (B) bound to DNA. **M13** and **M18** concentration was 0.2 μM. Shown in insets are the representative binding isotherms of dyes.

It appeared that the site exclusion parameter, i.e., number of DNA units excluded by the cyanine molecule and molar fluorescence (a) have the values typical for DNA-dye complexation. The association constants were found to have the magnitude of 10^5 M^{-1}, suggesting that the dyes under study form stable complexes with DNA. The highest values of association constant were observed for **M13** and **M16**, whereas the molar fluorescence was the largest in the case of **M4**. To explain the observed findings, it should be noted that the binding constant reflects the strength of binding, whereas the term "molar fluorescence" in generally determines the fluorophore ability to respond to the changes in its microenvironment.

Table 4. Parameters of cyanine-DNA binding calculated using McGhee & von Hippel model

Dye	$K, \times 10^5, M^{-1}$	n	$a, \times 10^3, M^{-1}$
M1	0.7±0.2	2.5±0.7	1.2±0.4
M2	2.1±0.6	1.5±0.4	0.2±0.04
M3	0.5±0.1	1.9±0.5	1.2±0.3
M4	0.2±0.07	1.0±0.3	8.4±2.5
M5	2.0±0.6	1.5±0.5	0.4±0.1
M6	0.5±0.1	1.0±0.3	8.0±2.4
M7	1.2±0.4	2.5±0.7	1.3±0.4
M8	2.3±0.7	3.2±0.9	1.4±0.4
M9	0.5±0.1	2.9±0.8	5.2±1.6
M10	0.9±0.2	3.0±0.9	1.3±0.4
M11	0.8±0.2	4.8±1.4	10.6±3.1
M12	1.2±0.4	3.2±0.9	1.8±0.5
M13	7.6±2.2	3.1±0.9	3.7±1.0
M14	0.6±0.2	3.1±0.9	1.6±0.5
M15	1.3±0.4	1.6±0.4	2.6±0.8
M16	5.7±1.1	1.9±0.6	1.9±0.5
M17	3.3±0.8	1.4±0.4	3.9±1.2
M18	3.1±0.5	1.7±0.5	1.3±0.4
M19	4.0±0.3	1.5±0.4	1.0±0.3

In the following, it is tempting to analyse the molecular events underlying the complex formation between DNA and cyanines. The relative affinity of the dyes for nucleic acids is largely determined by the nature of their association. Generally, main modes of interactions between guest molecules and DNA occur via: i) intercalation, ii) minor groove binding, and/or iii) major groove binding. While the latter mode is specific mostly for large macromolecules such as proteins, intercalation and minor groove binding represent the most common pathways of DNA complexation with small molecules. Intercalation is typical for planar aromatic cationic molecules. This type of binding results from incorporation of dye planar aromatic moiety between DNA base pairs following by unwinding and lengthening of DNA helix. In contrast, groove binders are crescent-shaped heteroaromatic structures possessing conformational flexibility which allows the dye molecule to adjust within the DNA groove. Compared to intercalation, groove binding distorts native DNA conformation to the less extent. Unfortunately, these two modes of dye-DNA interactions are spectroscopically indistinguishable. However, detailed analysis of the available literature on DNA interactions with fluorophores allowed us to suggest that preferential mode of cyanine binding to DNA in our case is intercalation. This assumption was based on the following considerations. Association constants for intercalators that bind to DNA by hydrophobic, van der Waals and electrostatic forces do not exceed $10^6 \, M^{-1}$, whereas the K value observed for DNA complexes of groove binders, especially those stabilized by hydrogen bonds (netropsin), may even be larger than $10^8 \, M^{-1}$ [81, 82]. For example, K values for ethidium bromide, acridine orange and Thiazole Orange, well-known intercalating agents, are equal to $1.5 \times 10^5 \, M^{-1}$, $2.69 \times 10^4 \, M^{-1}$, and $10^6 \, M^{-1}$, respectively [82-84]. The dye Cyan 2 has association constant even smaller, $\sim 10^4 \, M^{-1}$ [82]. The binding constants for asymmetric Thiazole Orange derivatives reported in [62] fall into the range $(0.66-7.58) \times 10^5 \, M^{-1}$ suggesting the intercalation binding mode. Other monomethine dyes (TOTO, YO, YOYO, etc.), the planar aromatic moiety of which incorporates between the DNA base pairs, have the binding constants of the order $10^6 - 10^7 \, M^{-1}$ [85].

The assumption about intercalating binding mode of cyanine dyes is confirmed also by the fact, that the size of the examined dyes does not exceed 2.2 nm, that is comparable with the space available for intercalation between the DNA base pairs (the diameter of dsDNA is ~2 nm). In addition, the groove binders are generally crescent-shaped structures possessing conformational flexibility, that is more usual for polymethines, while the dyes under study belong to the monomethine cyanine group. Finally, according to the principle of the nearest neighbor exclusion, the binding of one intercalating molecule between two base pairs hinders access of the next binding site to another intercalator [73], so the highest possible dye-base pair ratio for intercalation is 1:2. For groove binding agents one dye molecule occupies at least 3-5 base pairs which is a consequence of more elongated and ribbon like fluorophore structure. As shown in Table 3, the obtained estimates of site exclusion parameter are about 2, lending additional support to the assumption about the intercalation binding mode. Exceptions are only **M8**, **M11-M14**, for which site exclusion parameter attains the values ranging from 3 to 4.8. To determine whether it is indicative of another binding mode or simple mathematical artifact, additional experiments are further needed.

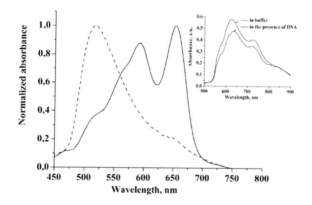

Figure 5. Normalized UV-visible absorption spectra of **P2** in buffer (solid line) and in the presence of DNA (dashed line). **P2** concentration was 3.5 µM. DNA concentration was 4.4 µM. Shown in inset are absorption spectra of **H2** in buffer and in the presence of DNA. **H2** concentration was 10 µM. DNA concentration was 34 µM.

Modulating Effect of the DNA on Aggregation Behaviour of Heptamethine and Pentamethine Dyes

Further, the applicability of heptamethine and pentamethine cyanine dyes for the DNA recognition has also been studied. Association of pentamethine cyanine dyes (except for **P5**) with DNA resulted in attenuation of the monomer absorption band (Figure 5, Table 5) and an appearance of a second blue-shifted band with respect to the monomeric one. The magnitude of this effect was found to depend on the dye structure. For **P1**, **P2**, **P3** and **P4** the monomer peak almost disappeared and the shift of the aggregate absorption towards the lower wavelengths was observed. Furthermore, this blue-shifted band was found to increase with the DNA concentration clearly indicating the binding of pentamethines to the double stranded DNA. These findings suggest the enhancement of the pentamethine dye aggregation (except for **P5**) in the presence of DNA. It should be noted, that penthamethine dyes **P6** and **P7** also aggregate readily on the DNA template, however the aggregation is less favorable than that of **P1**, **P2**, **P3** and **P4** on the basis of the fact that the residual monomer band is stronger than the H-aggregate band.

Table 5. Absorption maxima of pentamethine and heptamethine cyanine dyes in the presence of DNA

Dye	$^*\lambda_A^M$	λ_A^H
P1	640	504[a]
P2	654	518[a]
P3	642	585, 539[a]
P4	655	554[a]
P5	653	542
P6	648	597
P7	647	592, 540
H1	831	591[a]
H2	810	624[a], 729

$^*\lambda_A^H$, λ_A^M are absorption wavelengths corresponding to the maximum of the spectra, of aggregated and monomeric dye, respectively;

[a]The most intensive maximum, in the case that this maximum does not belong to the dye monomers.

The role of DNA as a template for cyanine aggregate growth was also reported in other works [48, 86, 87]. It was postulated that minor groove of nucleic acid provides the favorable environment for cyanine binding and self-assembly. To exemplify, Norden and Tjerneld reported from the circular dichroism studies that pseudoisocyanine formed aggregates on DNA template [88]. Dye aggregation on the DNA scaffold was also observed for bichromophoric dye K-6 and trichromophoric probe K-T [89]. In both cases the authors demonsrated that aggregates formed on the DNA template have the structure distinct from that observed in the aqueous solution [88, 89]. Armitage et al. suggested that the DNA-templated aggregation of cyanines is characterized by the two-level cooperativity

[48]. At the first one the association of one monomer molecule in the DNA groove facilitates the binding of a second monomer. The origin of this effect lies in unfavorable van der Waals interactions between the planar dye and nonplanar DNA. Thus, the association with another monomer and formation of the dimer will be more optimal. At the second level, formed dimer stimulates accumulation of other dimers. Adjacent dimers within the minor groove can interact in an end-to-end fashion. The helicity of the DNA template is imparted to the dye aggregate, resulting in the induced chirality. This phenomenon may be explained as follows. Formation of the dye dimer in DNA groove is known to widen the groove. Such preopening will induce the association of additional dye dimers since binding to the already perturbed groove will require less energy penalty compared to unperturbed one.

The complexation of heptamethine cyanine dyes and penthamethine **P5** (inset of Figure 5) with DNA resulted in attenuation of the intensity due to dilution of the sample concentration, and almost no significant change in the shape of the spectra. It should be noted, that H-aggregates represent the dominating fraction of cyanine dyes **H1**, **H2** and **P5** in the buffer solution. It seems that face-to-face contacts between dye molecules are stronger than binding strength of heptamethines and **P6** with DNA, whereas the binding of cyanine aggregates of higher order to DNA is blocked by the width of the minor groove of DNA.

Binding of Cyanine Dyes to the Lipid Vesicles

Cyanine dyes have been widely employed to trace the processes occurring in biological assemblies such as micelles and vesicles, since their photophysical behavior depends strongly on the properties of the surrounding medium [43]. Therefore, to gain further insights into the interactions of cyanine dyes with biological objects, the behavior of penthamethine and monomethine dyes in the lipid environment have been studied.

Binding of Heptamethine Cyanine Dyes to the Model Lipid Membranes

Presented in Figure 6 are the absorption spectra of the studied dyes in the presence of liposomes. The association of heptamethines with PC lipid vesicles led to the decrease of the absorption peak, attributed to the H-aggregates and an appearance of two bands, characteristic of the H-dimer and monomer absorption. The position of H-dimer peak was around 713-720 nm, while monomers were the most absorptive at 819-826 nm, respectively. These findings suggest that upon transition to the lipid environment highly organized plane-to-plane arrangement is disrupted, thereby favoring the randomized monomer arrangement. The results presented in Figure 6 indicate that partition of heptamethine cyanine dyes into the liposome is two-stage cooperative process. Initially, only H-aggregates are present. At the first stage, the addition of the liposomes in relatively low concentrations stimulates the dimer formation. The observed dimerization is most probably arises from the decrease of polarity of the aggregates environment.

It seems very likely that at low-polar solvents the dispersion forces become incapable to govern the staking orientation of the molecules in H-aggregates, so we observed an accumulation of dimers in the polar region of lipid bilayer. At the second stage, increase in lipid concentration promotes the appearance of monomeric dye form, and, finally, the contribution of monomeric band in the total spectrum prevails over the other dye forms. The driving force for this process involves most probably hydrophobic interactions. It may be supposed that for hydrophobic moieties at the heterocyclic ring of heptamethines dyes will be energetically more favorable to penetrate deeper to the lipid bilayer orienting parallel to the lipid acyl chains. These hydrophobic interactions led to further decrease in the polarity of lipid dye microenvironment, since it is known that the polarity profile is change dramatically from polar to hydrophobic core of liposomes [90].

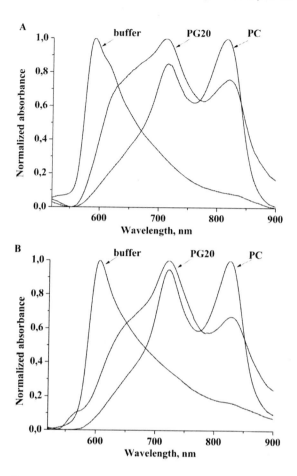

Figure 6. Normalized UV-visible absorption spectra of **H1** (A) and **H2** (B) in buffer, in the presence of pure PC and PG-containing liposomes. **H1** and **H2** concentrations were 12.2 µM and 13.5 µM, respectively. Lipid concentration was 8.9 µM.

Interesting, the magnitude of the dissociation of **H1** and **H2** is considerably smaller in the presence of PG-containing liposomes in comparison with pure PC bilayer. In the presence of negatively charged PG the concentration of H-aggregates free in buffer solution decreases with an increase in lipid concentration and an appearance of dimeric and monomeric dye species. However, in the PG bilayer at the same lipid concentration the contribution of dimeric band in the total spectrum was prevailing over the other dye forms, indicating about the slower degree of disaggregation. The fact that the decrease of liposome-induced cyanine aggregation was more pronounced in the neutral PC bilayers compared with negatively charged PG-containing vesicles, suggests that the disruption of highly organized molecular arrangements is electrostatically-controlled. It may be supposed that strong electrostatic lipid-dye interactions with PG vesicles ensure the lipid bilayer surface association of the cyanines, increasing thereby their local concentration and eventually hampering the dissociation process. Moreover, we should not exclude the fact that the physicochemical properties of bilayer could also affect on lipid-induced disruption of ordered nanoassemblies of noncovalently coupled probes. Therefore, to interpret the observed differences in disaggregation tendency values of heptamethines the influence of PG on lipid bilayer structure should be considered.

Recent studies revealed an increase in bilayer hydration in the presence of PG [91, 92]. In particular it was found a significantly higher hydration of carbonyl groups in the PG-containing membranes compared with pure PC lipid bilayer [91, 93]. Moreover, an additional argument in favor of the increase in the degree of bilayer hydration in the presence of anionic lipids comes from the spectral response of a number of environment-sensitive fluorophores [94-96]. It was shown that the changes in hydration extent may considerably affect molecular organization of a lipid bilayer. In particular, increase of water content in headgroup region was reported to modify the alignment of choline-phosphate dipole and lateral packing of hydrocarbon chains [97]. Moreover, as follows from ^{2}H-NMR spectroscopic studies, the presence of PG in PC bilayers alters the orientation and mobility of adjacent choline headgroups [97, 99]. Likewise, it was found that PG induces a much dense packing of chain atoms in the near-the interface region of hydrophobic core of the mixed bilayers compared with pure PC [100] and exerts a stabilizing influence on zwitterionic bilayers [101]. All these rationales let us to suppose that the ability of PG to change the conformation and dynamic of choline headgroup along with their tendency to form more stable and condensed bilayers could affect the molecular organization of model membranes in such a manner, that the second level of dissagregation becomes thermodynamically unfavorable for heptamethines, so in the negatively charged membranes these dyes accumulate in the polar membrane region presumably in the form of H-dimers.

Binding of Pentamethine Cyanine Dyes to the Model Lipid Membranes

Figure 7 illustrates the representative absorption spectra of penthamethine cyanine dyes in the absence and presence of CL67 lipid vesicles. Binding of pentamethine cyanine dyes **P3**, **P4**, **P6** and **P7** to the model lipid membranes was followed by a decrease in the monomer absorbance. The absorbance spectrum changed from 'two-peak' to 'three-peak' form manifesting itself in the appearance of the third well-defined hypsochromic band, suggesting that in the lipid-bound form these dyes are represented by monomers, H-dimers and H-aggregates (Table 6). The magnitude of the absorbance decrease was more pronounced for the monomeric species, resulting in the nearly equal contribution of different dye fractions into the overall spectrum. Moreover, the monomeric band of **P1** and **P2** exhibited significant decrease upon the formation of dye-lipid complexes, and the contribution of the aggregated species dominated over the monomeric ones. This implies that H-aggregates are the dominating species of **P1** and **P2** in lipid-bound form. In turn, the association of **P5** with the lipid vesicles led to the diminishing of the peak, attributed to the dye H-aggregates (540 nm) and appearance of the new peak around 660 nm, describing the absorbance of the monomers. The observed effect assumes that dye-lipid complexation breaks **P5** aggregates, and the dye experienced transition into the monomeric state.

Taken together, these results suggest that transfer of the cyanine dyes into the lipid environment enhances their aggregation propensity. Exception was only the dye **P5** which showed the opposite behavior. The aggregation of cyanine dyes is well-studied process. In aqueous solution the formation of cyanines occurs at high fluorophore concentrations. The situation changes when the dye is transferred to the low-polarity environment where cyanines self-associate easily even in diluted solutions at room temperature [34]. The molecular basis for such phenomenon lies in the fact that aggregation of the dyes in the low polar solvents, in contrast to highly polar solvents, is controlled not only by the dispersion forces (which govern the stacking orientation of the dye monomers in the aggregate structure), but also by electrostatic interactions [34]. It was postulated that in solvents with high dielectric constant cyanines exist in the form of solvated ions [102]. Upon decreasing the values of dielectric constant, cyanine dyes change their functional state to so-called 'contact ionic pairs'. Further lowering the dielectric constant results in close approaching of these pairs with their concomitant association into the aggregates, stabilized additionally by electrostatic interpair forces.

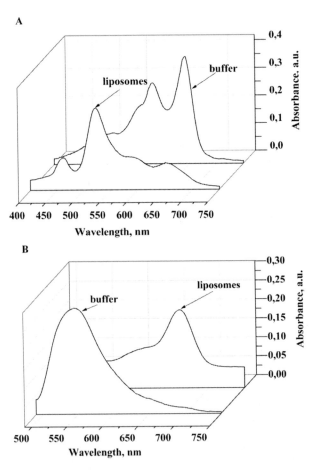

Figure 7. Representative absorption spectra of **P1** (A) and **P5** (B) recorded in the presence of liposomes. **P1** and **P6** concentration were 4.5 µM and 4.3 µM, respectively. Lipid concentration was 81,9 µM.

In the case of our systems, it is likely that lipid vesicles not only provide the environment with reduced polarity, but also serve as the template for dye oligomerization. It may be supposed that strong electrostatic lipid-dye interactions ensure the lipid bilayer surface association of the cyanines, increasing thereby their local concentration and promoting eventually formation of stacked H-aggregates.

Table 6. Absorption maxima of pentamethine cyanine dyes in the presence of liposomes

Dye	λ_A^H	$*\lambda_A^M$
P1	583, 521[a]	649
P2	531[a], 480	644
P3	548, 594[a]	652
P4	604[a], 563	664
P5	542	660
P6	603[a]	660
P7	601, 555	654

$*\lambda_A^H$, λ_A^M are absorption wavelength of aggregates maxima and monomeric dye form;
[a] The most intensive maximum, in the case that this maximum does not belong to the dye monomers.

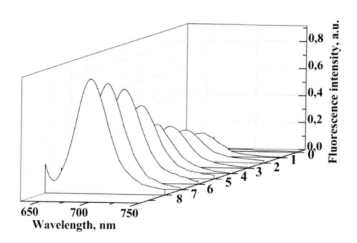

Figure 8. Emission spectra of P5 in CL67 lipid bilayer. **P5** concentration was 1 μM. Lipid concentrations were 0; 8.2 (1), 16.2 (2), 24.1 (3), 32 (4), 47.6 (5), 78.7 (6), 118 (7) and 157,5 μM, respectively.

In contrast, the results obtained with **P5** suggest, that the aggregation propensity of this dye reduces upon its incorporation into the lipid membranes. It seems likely that in more hydrophobic medium such as lipid bilayer, highly organized plane-to-plane arrangement, which is characterized by a broad H-band, is distrupted, giving rise to the formation of randomized monomer arrangement. Additional arguments in favor of the dye disaggregation come from an appearance of the fluorescence spectra for **P5** in the range 630-750 nm with the maximum at 675 nm (Figure 8). It is generally known, that H-aggregates are nonfluorescent in nature. This fact represents the spectroscopic signature of the aggregate formation according to the excitation coupling theory [37]. Therefore, **P5** dye is advantageous as a membrane probe since it background fluorescence from the buffer can be neglected. The high fluorescence enchancement of **P6** in the presence of liposomes indicates that probe oligomerization in water is likely to be reversible. Moreover, with increasing lipid concentration the fluorescence intensity achieved saturation at probe/lipid ratio 1/157, which is comparable with other membrane probes [103, 104].

Association of Cyanine Dyes with Bovine Serum Albumin

Cyanine dyes have been successfully employed for the protein detection and characterization. To exemplify, cyanine dyes were found to form strong complexes with the bovine and human serum albumins [57, 105, 106]. Moreover, binding of meso-substituted anionic thiacarbocyanines to the human serum albumin results in cis-trans isomerization and, as a consequence, the appearence of the dye fluorescence [106]. Therefore, at the next step of our study, the spectral properties of the examined pentamethine and heptamethine cyanine dyes were analyzed upon their binding to bovine serum albumin. Absorption spectra of **P3**, **P5**, **P6** and **P7** in the presence of the protein were featured by two main bands corresponding to monomers and H-aggregates (Table 7).

Table 7. Absorption maxima of pentamethine cyanine dyes in the presence of bovine serum albumin

Dye	λ_A^H	*λ_A^M
P1	589, 543	651
P2	543[a], 502	651
P3	586	640
P4	597, 566	655
P5	542	653
P6	600	650
P7	592	647

* λ_A^H, λ_A^M are the absorption wavelength of aggregates maxima and monomeric dye form;

[a]The most intensive maximum, in the case that this maximum does not belong to the dye monomers.

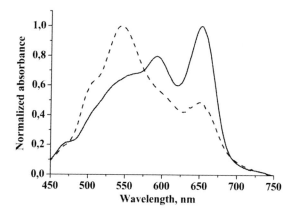

Figure 9. Normalized UV-visible absorption spectra of P2 in buffer (solid line) and in the presence of BSA (dashed line). P2 concentration was 3.5 µM. DNA concentration was 4.4 µM.

The position of monomer peak was around 640-650 nm while aggregates were the most absorptive at 586, 542, 600 and 592 nm, respectively. Furthermore, the BSA addition to the dye solution resulted in 1.3, 1.2, 1.2 and 1.1-fold decrease of the monomer and aggregate absorption, respectively, without the modification in the shape of the spectrum. In turn, the association of **P1** and **P4** with BSA was followed by 1.9 and 2.3-fold reduction of the monomer absorbance, respectively, and 1.2-fold decrease in the aggregate absorption. In addition, the formation of

BSA complexes with **P1** and **P4** led to the appearance of the third blue-shifted shoulder centered at 543 and 566 nm, respectively, reflecting the formation of H-aggregates of higher order. Notably, the absorption of aggregate bands was slightly lower compared to the monomeric one, suggesting that in the presence of BSA the concentration of the dye monomers decreases and the concentration of aggregates increases, but the contribution of these fractions into the overall spectral response is virtually equal.

An interesting behavior upon formation of the dye-protein complexes was revealed only for **P2** (Figure 5). In the absence of BSA the absorption spectra of this dye has two main bands at 654 (monomers) and 592 nm (H-dimers) coupled with a shoulder at 543 nm (H-aggregates). Binding of BSA provoked 2.3-fold drop in the monomeric absorption, almost total diminishing of the dimeric band and 1.2- fold rise in H-aggregate band. Moreover, the contribution of H-aggregate band into the overall spectrum was prevailing over the other dye forms. This observation mirrors the ability of BSA to increase the H-aggregation propensity of **P2**, resulting in the domination of the dye aggregated fraction in protein-bound state.

The binding of heptamethine cyanine dyes was monitored by evaluating the changes in the absorbance at fixed dye concentration with increasing the protein concentration (Figure 10). The changes in the absorption spectrum at the presence of protein can be summarized as follows:

I. The association of heptamethines with BSA led to the diminishing of the peak, attributed to the dye H-aggregates (588 and 634 nm for **H1** and **H2**, respectively) and appearance of the new peak around 811 and 818 nm, characteristic of the monomer absorption. The observed effect assumes that dye-protein complexation breaks the heptamethine aggregates, and the dyes undergo the transition into the monomeric state. However, the contribution of H-aggregate band into the total spectrum was still prevailing over the other dye forms.

II. The reduction of H-aggregate peak is accompanied by a red-shift of the absorption maximum from 588 nm to 645 nm and from 633 nm to 658 nm for **H1** and **H2**, respectively.

III. The formation of protein-dye complexes resulted in the broadening of the H-aggregates absorption spectrum, coupled with appearance of a shoulder around 708-715 nm.

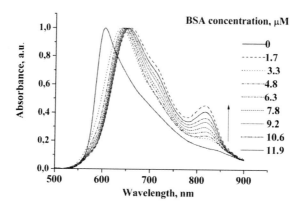

Figure 10. Normalized UV-visible absorption spectra of **H2** in the presence of increasing concentrations of BSA. **H2** concentration was 13.5 μM.

Together with the absence of isosbestic point between aggregated and monomer dye forms, these findings indicate that the heptamethine-protein interaction leads to the concomitant rupture of H-aggregate in buffer solution with the subsequent formation of dimeric and monomeric dye molecules. Most probably, the binding of heptamethines to BSA was strong enough to cause the dissociation of aggregates with aggregation tendency highly reduced after the formation of the BSA-dye complexes. The distruption of plane-to-plane arrangements followed by the formation of randomized monomers in the presence of BSA was reported also for merocyanine 540 [107], tricarbocyanine and heptamethine dyes [41, 108] and was interpreted as arising from higher strength of interactions between the dye and the BSA compared to the coupling between the cyanine monomeric species within the aggregate. The possible explanation of disassembly of H-aggregates in the presence of BSA in our study could lie in the strong electrostatic interaction between cyanine dyes and protein.

The isoelectric point of BSA is ~4.7 [109, 110], so at pH 7.4 this protein bears a negative electric charge. It can be seen from the structure of **H1** and **H2** that these dyes are cationic in nature, therefore the binding of heptamethines to BSA is likely to involve the electrostatic attraction.

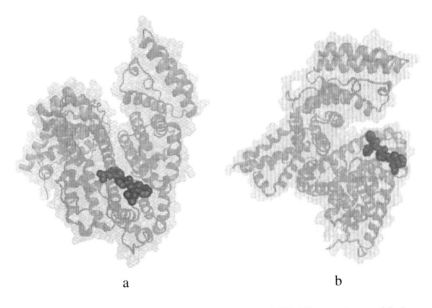

Figure 11. Schematic representation of the **H1** (A) and **H2** (B) complexes with the bovine serum albumin, obtained using PatchDock/FireDock servers.

In addition, crystal structure of BSA shows that the principal regions of ligand-binding sites are located in hydrophobic cavities in bovine serum albumin subdomains, referred to as the sites I and site II according to terminology proposed by Sudlow et al. [110]. Since heptamethine dyes under study contain hydrophobic groups in their structure, we can hypothesize that hydrophobic interactions along with electrostatic attraction serve as driving forces for dye transfer from aggregated to monomeric form. Previously, Patonay and coworkers characterized the hydrophobic binding pockets of the albumin molecule and observed that the binding affinity of cyanine dyes is attributed to the substitution of hydrophobic moieties at the heterocyclic ring of the cyanine dyes [53]. Numerous studies indicate that cyanine dyes generally exhibit a high specificity to the albumin binding sites I and II [41, 111], so one can hypothesize that similar mode of interactions is realized in our systems.

In order to further identify the preferred bovine serum albumin binding sites of heptamethine cyanine dyes, as well as the nature of interactions involved in the dye-protein complexation, a series of simple docking studies were performed. As seen in Figure 11, heptamethine dye **H1** binds preferentially to the subdomain IIA (Sudlow's site I), whereas **H2** molecules seem to occupy the subdomain IB. Furthermore, **H1** molecules are surrounded by the hydrophobic (Ala-290, Ala-341, Leu-197, Pro-338, Trp 213, Pro-446), charged (Arg-217, Arg-194, Lys-294, Asp450, Glu-291, Glu-293, Glu-339) and polar (Tyr-451, Ser-453, Ser-442, Tyr-451) amino acid residues. The charged residues in close proximity to the heptamethine dye H2 are Arg-185, Lys-114, Lys-136, Glu-182, Glu-140. Moreover, H2 is associated with hydrophobic residues Leu-115, Leu-122, Leu-178, Phe-164, Phe-133, Pro-117. Thus, the molecular docking results reflect that both hydrophobic and electrostatic interactions play a substantial role in the association of **H1** and **H2** with bovine serum albumin.

Interaction of Cyanine Dyes with Native and Fibrillar Lysozyme

Cyanine dyes have been successfully employed as non-covalent labels for the detection of the specific type of protein aggregates – amyloid fibrils, whose formation is involved in the pathogenesis of Alzheimer's disease, systemic amyloidosis, etc. [112]. To exemplify, pinocyanol dye appeared to be more sensitive to Aβ-fibrils than Congo Red [113], trimethine cyanines T-49 and SH-516 exhibited substantial fluorescence increases in the presence of fibrillar beta-lactoglobulin and α-synuclein [114], while the novel thiacarbocyanines [30], trimethine cyanine dye 7519 [115] and carbazole-based cyanines [116] were used as anti-amyloid drugs. Notably, the ability of the dyes to inhibit amyloid fibril formation is connected with their tendency to form aggregates in solution due to strong stacking interactions with the aromatic amino acids, as was shown for phtalocyanines, Congo Red, etc. [117, 31]. In view of this, on the next step of our study we ascertained the ability of heptamethine cyanine dyes to detect fibrillar lysozyme. More specifically, the quantitative characterization of the interactions between H1, H2 and native/fibrillar lysozyme was performed taking into account the dye monomer-aggregate equilibria in the buffer [51].

The absorption spectrum of **H1** in aqueous solution can be decomposed into at least 4 different components, corresponding, presumably, to the dye monomers (M, λ_{max}~808 nm), H-dimers (D, λ_{max}~700 nm) and higher order H-aggregates (H1, λ_{max}~ 595 and H2, λ_{max}~ 640 nm). The absorption maximum of the heptamethine dye **H2** for H1, H2, D and M are about 595 nm, 630 nm, 695 and 815 nm, respectively. The bis-heptamethine cyanine dyes have been reported to form predominantly H-dimers in aqueous solution [68]. The formation of H-aggregates in buffer starts at the lower probe concentrations [118] as compared to those (*ca.* 10^{-5}–10^{-3} M^{-1}) required for the dye J-aggregation [119]. Notably, some amount of **H1** and **H2** J-aggregates could appear in buffer, as judged from the broadening of the monomer absorption bands [42, 120]. Indeed, the incorporation of positively charged or bulky groups to the dye molecule resulted in the enhancement of J-aggregation [121, 122].

Numerous approaches have been employed to describe qualitatively cyanine aggregation in aqueous solutions. To exemplify, Taticolov and coworkers have estimated the aggregation constant and the number of the anionic cyanine dyes from the double-logarithmic plots of the J-aggregate absorbance versus the monomer absorbance [22]. Analysis of the average value of the extinction coefficient of naphtalocyanines observed at a certain wavelength, yielded n ~1.5–2.0, and K_{agg} ~ $4.5 \cdot 10^{-3}$–$3.1 \cdot 10^{-2}$ μM^{-1} [123]. Assuming the simplest case, i.e., the monomer-dimer equilibrium, the dimerization constants were estimated by a number of graphical methods, revealing the values *ca.* $(3.8–8.5) \cdot 10^{-3}$ μM^{-1} for benzindocarbocyanine dyes [124]; *ca.* $2.6 \cdot 10^{-3}$ μM^{-1} for mononuclear cobalt phthalocyanines [42]; *ca.* $(7.4–110) \cdot 10^{-3}$ μM^{-1} and n ~ 1.2–1.7 for thiacyanines [121]; and *ca.* 7.4–9.3 for the tetrasulfonated phtalocyanines

[123], etc. In a general case, for the monomer–n-aggregate equilibria, the numerical solution of the equation, which describes the balance between the total probe concentration and the amounts of the monomer, dimer, .., n-aggregate species, could be found, revealing the concentration of the monomer species in buffer and the averaged aggregation constant $K_n^{tot} = K_1 \cdot K_2 \cdots K_n$, where $K_1, K_2, .., K_n$ are the constants of the dye dimerization, trimerization,..., n-merization, respectively. This approach was employed, for instance, by Schutte and coworkers for the octalosubstituted phtalocyanine, which the values of $K_1 = 1.5$ μM^{-1} and $K_1 \cdot K_2 = 0.08$ μM^{-2} were obtained [124].

Therefore, in order to characterize the processes occurring in the dye-protein system we first directed our efforts towards estimation of the equilibrium constants for the H-dimer (K_d) and H-aggregate (K_h) formation, and the averaged dye aggregation number (n) in buffer, assuming the existence of the dye monomer-H-dimer and dye-H-aggregate equilibria [22]:

$$nC_m \rightleftarrows C_h, K_h = \frac{C_h}{C_m^{\ n}} \tag{2}$$

$$2C_m \rightleftarrows C_d, K_d = \frac{C_d}{C_m^{\ 2}} \tag{3}$$

$$C_m + 2K_d C_m^{\ 2} + nK_h C_m^{\ n} = C_0, \tag{4}$$

where C_m, C_d, C_h, C_o are the concentrations of the dye monomers, H-dimers, H-aggregates, and the total dye concentration, respectively. Numerical solution of eq. (4) performed by simultaneous analysis of the absorption spectra measured at varying C_o, yielded rather realistic values of the aggregation number and equilibrium constants, viz. { $K_h = 85$ μM^{-1}

$K_d = 7 \mu M^{-1}$, $n = 4$ } and { $K_h = 120 \mu M^{-1}$, $K_d = 5 \mu M^{-1}$, $n = 4$ } for **H1** and **H2**, respectively, the values being similar to those reported for cyanines elsewhere [22, 122].

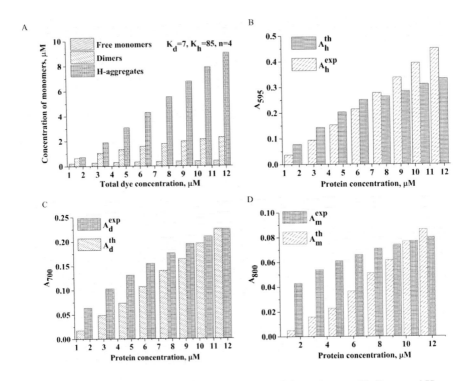

Figure 12. Theoretically predicted concentrations of the monomer, H-dimer and H-aggregate dye species at different total concentrations of **H1** in buffer (A), the measured (taken from the deconvoluted dye spectra) and calculated absorbances of the H-aggregates (B), H-dimers (C) and monomers (D).

The recovered parameters were further used to calculate the dye monomer, H-dimer and H-aggregate concentrations, plotted as a function of the total dye concentration in Figures 12A and 13A, respectively. It appeared that amount of the dye monomers in the H-aggregate and free monomer species were the highest and the lowest, respectively, within the examined C_o range. The observed trend is in good agreement with the measured band intensities of the dye species in buffer. Furthermore, the discrepancy between the measured and calculated absorbances of **H1/H2** monomers, H-dimers and H-aggregates (the values were taken from the deconvoluted dye spectra) was found to be the lowest for $\varepsilon_d = 0.1 / 0.197$ $\mu M^{-1} cm^{-1}$, $\varepsilon_h = 0.05 / 0.1 \mu M^{-1} cm^{-1}$, where ε_d and ε_h are the extinction coefficients of the dye monomers in the H-dimers and H-aggregates, respectively.

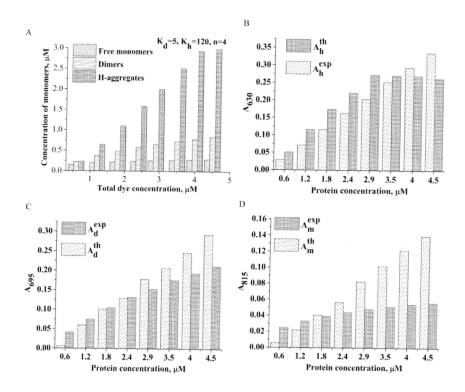

Figure 13. Theoretically predicted concentrations of the monomer, H-dimer and H-aggregate dye species at different total concentrations of **H2** in buffer (A), the measured (taken from the deconvoluted dye spectra) and calculated absorbances of the H-aggregates (B), H-dimers (C) and monomers (D).

Figures 12B–D and 13B–D represent the measured and calculated absorbances at the wavelengths, corresponding to the monomer, dimer and H-aggregate dye species, over the overall dye concentration range. Our results agree with the fact that extinction coefficients of the dye monomers in the H-dimer and higher order H-aggregates are lower than that of monomer species [68, 120]. The value of K_h for dye **H2** exceeds that of **H1** by ~30%, presumably due to stronger dispersion forces between the two **H2** molecules [125]. Indeed, the structure of **H2** dimers should be more planar than that of AK7-5 due to the stronger stacking interactions between the monomer species. The recovered parameters, characterizing the dye monomer-aggregates equilibria in buffer, were further employed to describe the spectral behavior of the examined cyanines in the presence of native and fibrillar lysozyme.

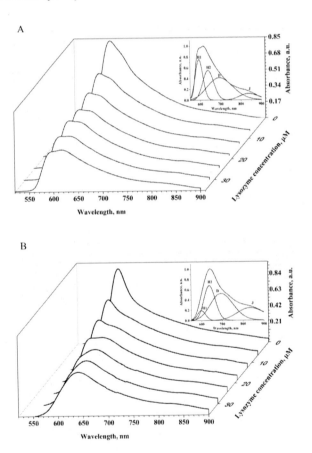

Figure 14. Absorption spectra of **H1** (A) and **H2** (B) at an increasing concentration of the native lysozyme (LzN). Normalized absorption spectra of **H1** (insert A) and **H2** (insert B) in the presence of 35.3 µM and 35.7 µM of the native lysozyme, respectively. **H1** and **H2** concentrations were 12.2 and 13.5 µM, respectively.

Heptamethine Cyanine Dye response to the Native Lysozyme

Upon addition of the native lysozyme to **H1** and **H2** in aqueous phase, the following effects have been observed: i) the marked decrease in the absorbance of H-aggregates, augmenting with increasing the protein concentration (Figure 14A, B); ii) the rise in the relative contribution of the H2 band to the overall spectrum; iii) slight absorbance increase at ~840 nm (55% and 26% at the maximum protein concentration, respectively).

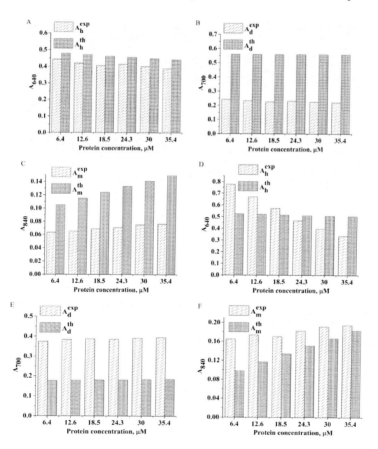

Figure 15. The measured (Am_exp) and theoretically predicted (Am_th) absorbances of the **H1/H2** H-aggregates (A,D), H-dimers (B,E) and monomers (C,F) in the presence of the native lysozyme calculated using the following sets of parameters:

$$\{K_b = 0.02 \mu M^{-1}, m=1, K_h = 85, K_d = 7\mu M^{-1}, n=4\} / \{K_b = 0.04\mu M^{-1}, m=1, K_h = 120, K_d = 5\mu M^{-1}, n=4\}$$

While considering the dye-protein interactions, it is important to find out what kinds of the dye species participate in this process.

Hypothesis 1. First, we supposed that the binding of the dye monomers to native lysozyme results in the shift of aggregation equilibria, i.e., disruption of the dye H-aggregates (Figure 15). Indeed, both the dye-protein complexes and the dye aggregates are stabilized by the same types of intermolecular interactions (van der Waals, H-bonding, electrostatic and hydrophobic) [120], and the dye disaggregation could occur as a result of the competition between the dye-protein and dye-dye binding [53, 119]. The disruption of cyanine H-aggregates has been observed in the presence of peptides and human serum albumin, indicating that the dye-protein complexation is more energetically favourable than the aggregate formation [68, 126]. Notably, monomer species of phtalocyanine dyes have been found to interact with native insulin, with the dye low-order self-associates playing a minor role in this process [127]. To test the validity of this hypothesis, an additional relationship should be added to the above set of eqns. {2–4}:

$$C_m + P \rightleftarrows C_{mp}, K_b = \frac{C_{mp}}{C_m (mP - C_{mp})},$$ (5)

where C_{mp}, P, K_b, m are the concentration of the protein-bound monomers, lysozyme concentration, association constant and stoichiometry for the dye-protein binding, respectively. Notably, the C_{mp} addendum was included into eq. (4). By solving the equation set {4,5}, we made an attempt to reproduce the spectral behavior of the examined dyes in the presence of lysozyme through looking for the appropriate set of parameters $\{K_b, m\}$. However, we failed to obtain reasonable values of the above parameters capable of providing good agreement between the experiment and theory. For instance, for $\{K_b = 0.02 \ \mu M^{-1}, m = 1\}$ at the maximum protein concentration 35.3 μM the discrepancy between the experimental and calculated values of A_{640} for **H1** was about 18%, while the calculated values of A_{700} and A_{840} were greater than the experimental ones by the factors of ~2.5 and ~2, respectively. Similarly, for $\{ K_b = 0.04 \ \mu M^{-1}, \ m = 1 \}$ at the maximum protein concentration 35.7 μM, the discrepancy between experimental and calculated A_{600} for **H2** was about 33%, while calculated values of A_{700} and A_{840} were lower and greater than the experimental ones by the factors of ~2 and ~1.5, respectively. Figure 15 represents measured and calculated absorbances at 640/ 600, 700 and 820 nm over the overall protein concentration range.

The obtained discrepancies between the experiment and theory imply that the assumption about the dye H-aggregate disruption as a result of the dye monomer binding to native protein cannot per se explain the observed spectral effects. Notably, the extinction coefficients were taken as 0.05/ 0.1 (H-aggregates) and 0.1/0.197 (H-dimers) $\mu M^{-1}cm^{-1}$ for **H1** and **H2**, respectively, while ε_{840} was assumed to be similar for free and bound monomer species ($\varepsilon_{840} = 0.197 \ \mu M^{-1}cm^{-1}$). Indeed, the absorptivity values of the protein-bound dye monomers appeared to be close to those of the free dye species, as was shown, e.g., for Thioflavin T and Congo Red [128, 129].

Hypothesis 1a. In order to eliminate the above discrepancies between the experiment and theory, it was suggested that both the dye monomers and H-dimers could bind to native lysozyme, leading to the shift of aggregation equilibria. This requires the inclusion of an additional eqn. into the above set $\{2-5\}$:

$$C_d + P \rightleftarrows C_{dp}, K_{bd} = \frac{C_{dp}}{K_d C_m^2 (m_d P - C_{dp})}, \qquad (6)$$

where C_{dp} is the concentration of the lysozyme-bound H-dimers, K_{bd}, m_d are association constant and stoichiometry for the H-dimer-protein binding, respectively. Notably, the C_{mp} and $2 \cdot C_{dp}$ addenda were included into eq. (4), where $2 \cdot C_{dp}$ is the concentration of the protein-bound monomers in dimeric form. Furthermore, eqns. 5 and 6 were combined under the assumption of different protein binding sites for the dye monomers and dimers. However, *Hypothesis 1a* did not lead to sufficient improvement of the theoretical binding curves. Specifically, for the set of **H1/H2** parameters $\{K_b = 0.02\,\mu\text{M}^{-1},\ m = 1, K_{bd} = 0.01\,\mu\text{M}^{-1},\ m_d = 1\}\ /\{K_b = 0.03\ \mu\text{M}^{-1},\ m = 1, K_{bd} = 0.01\,\mu\text{M}^{-1},\ m_d = 1\}$ the difference between the calculated A_{640}, A_{700} and A_{840} values were the same or greater than those obtained for the *Hypothesis 1*, indicating that **H1** and **H2** spectral behavior in the presence of native lysozyme is influenced to a minor extent by the dye monomer- and dimer-protein binding.

Hypothesis 2. In view of the above rationales, we fail to obtain reasonable values of providing good agreement between the experiment and theory, because the absorbance decrease in H-band (arising from the breaking of the dye H-aggregates upon the shift of the dye monomer–aggregate equilibrium) is presumably followed by the J-aggregate formation by the protein-bound dye monomers. This idea is corroborated by the lower absorptivity values and characteristic bathochromic shifts of lysozyme-bound **H1** and **H2** with respect to the monomer bands (Figure 15C, D) [113], and agrees with the previously reported formation of J-aggregates in the presence of BSA [130]. Finally, estimation of the half widths of the **H1/H2** monomer bands in DMSO at 808/815 nm, and the bands at 840 nm observed in the presence of 35 μM lysozyme, yielded the values $\Delta v_{1/2}$ ~50/60 nm and ~125/145 nm, respectively, suggesting the appearance of the new dye species upon the dye complexation with the protein [51, 120]. Taking into account the fact that J-aggregates possess narrow absorption bands, the broad band at 840 nm seems to be the superposition of the J-aggregate (of different size and molecular packing) and monomer bands. It is also noteworthy that higher stability of the dye **H1** H-aggregates and lower extent of J-aggregate formation as compared to **H2** (Figure 15 C,D) could be interpreted in terms of the different polarity of the dye binding sites [114]. Notably, an appropriate theoretical model, allowing quantitative characterization of the **H1**, **H2** complexation with native lysozyme, was not developed because J-aggregate absorption bands of **H1**, **H2** are very weak to resolve quantitative characterization of their formation. Furthermore, inclusion of a new equilibrium between the protein-bound monomers and J-aggregates to the set {2–6} will significantly complicate the solution of the obtained set of equations.

Modulation Effect of the Fibrillar Lysozyme on Aggregation Behaviour of Heptamethine Dyes

The changes produced by fibrillar lysozyme in the absorption spectra of **H1** and **H2** were found to differ from those caused by the native protein. As illustrated in Figure 16, the absorption maxima of H-aggregates are shifted by ~45 nm to longer wavelengths. This effect was accompanied by the absorbance decrease of **H1**, being less pronounced than that brought about by the native lysozyme (~7% at the maximum protein concentration relative to the lowest protein concentration 6.4 μM). Furthermore, the intensity of **H2** H-band was enhanced by a factor of ~1.4 at the maximum protein concentration. At the same time, a shoulder was appeared at 700 nm, reflecting the increase of the amount of H-dimer species (Figure 15). Likewise, the drop in the relative contribution of the aggreagate H1 band to the overall spectrum was much more pronounced for fibrillar lysozyme, compared to the native protein (Figure 14). Finally, marked absorbance increases were observed for **H1/H2** at 820 nm/830 nm (*ca.* ~7/2.5 times at the maximum protein concentration). The observed red shifts of the dye absorption maxima could result from both the decrease of the environmental polarity [131] and J-aggregate formation. For example, *ca.* 15–27 nm red shift of the monomeric absorption band was observed for pentamethine and thiacarbocyanine dyes bound to bovine serum albumin (BSA) and human serum albumin (HSA), respectively, relative to buffer [105]. Interestingly, the half-widths of the dye absorption bands at 820 nm/830 nm were $\Delta v_{1/2}$ ~85 nm and 90 nm, respectively. Although these values are close to the half-width of the monomer band, the appearance of some amount of J aggregates upon the dye incorporation into lysozyme fibrils cannot be excluded. Similarly, the binding of thiacarbocyanine 7514 to insulin amyloid fibrils resulted in the greatly increased absorbance of the band corresponding to the monomeric dye-fibril complex, and slightly enhanced intensity of the dye J-aggregate band [132]. It should be noted in this context that the self-stacked dimers of the cyanine dye YOYO-1 were disrupted in the presence of fibrillar Aβ(1–42) due to the self-stacking to non-stacking transition of the two YO moieties [133].

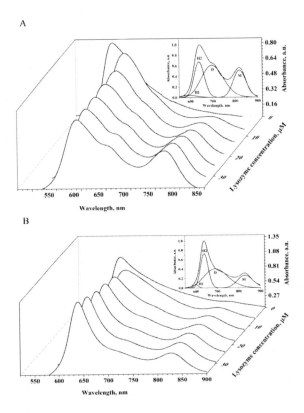

Figure 16. Absorption spectra of **H1** (A) and **H2** (B) at increasing concentration of the fibrillar lysozyme (LzF). Normalized absorption spectra of **H1** (insert A) and **H2** (insert B) in the presence of 35.3 μM (insert A) and 35.7 μM (insert B) of fibrillar lysozyme. The **H1** and **H2** concentrations were 12.2 and 13.5 μM, respectively.

Next, the same hypotheses as in the case of native lysozyme have been invoked to explain the spectral behavior of the examined cyanines in the presence of lysozyme fibrils.

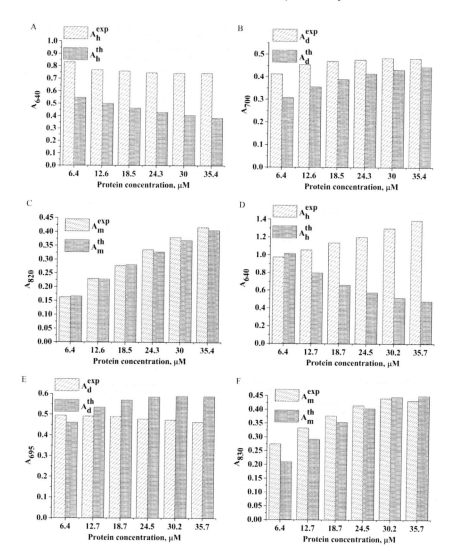

Figure 17. The measured (Am_exp) and theoretically predicted (Am_th) absorbances of the **H1/H2** H-aggregates (A and D), H-dimers (B and E) and monomers (C and F) in the presence of fibrillar lysozyme, calculated using the set of parameters:

$\{K_b = 0.6 \mu M^{-1}, m = 0.3, K_{bd} = 0.2 \mu M^{-1}, m_d = 0.8, K_h = 85, K_d = 7 \mu M^{-1}, n = 4\}$ /
$\{K_b = 0.8 \mu M^{-1}, m = 0.35, K_{bd} = 0.3 \mu M^{-1}, m_d = 2, K_h = 120, K_d = 5 \mu M^{-1}, n = 4\}$

Hypothesis 1. First, since the most substantial increases of the **H1** absorbance were observed at 820 nm, we found the set of parameters {

$K_b = 0.5~\mu\text{M}^{-1}$, $m = 0.3$, $\varepsilon_{820} = 0.197~\mu\text{M}^{-1}\text{cm}^{-1}$} providing good agreement between experiment and theory, with the difference between the experimental and calculated absorbances being less than 4% at the maximum protein concentration. On the contrary, calculated values of A_{640} ($\varepsilon_{640} = 0.05~\mu\text{M}^{-1}\text{cm}^{-1}$) and A_{700} ($\varepsilon_{640} = 0.1~\mu\text{M}^{-1}\text{cm}^{-1}$) were less than the experimental ones by 30.5% and 53%, respectively, at the maximum protein concentration. Similarly, for heptmethine dye **H2** the following parameter set { $K_b = 0.5~\mu\text{M}^{-1}$, $m = 0.3$, $\varepsilon_{830} = 0.197~\mu\text{M}^{-1}\text{cm}^{-1}$} provided the difference between the experimental and calculated absorbances at 830 nm less than 16% at the maximum protein concentration. The values of A_{640} ($\varepsilon_{640} = 0.1~\mu\text{M}^{-1}\text{cm}^{-1}$) and A_{700} ($\varepsilon_{640} = 0.197~\mu\text{M}^{-1}\text{cm}^{-1}$) were less than the experimental ones by 17% and 35%, respectively. Thus, the assumption that only the dye monomers are capable of associating with lysozyme fibrils cannot reproduce the behavior of the H-dimers and H-aggregates with sufficient accuracy.

Hypothesis 1a It cannot be excluded that the binding of both the dye monomers and H-dimers to the fibrillar lysozyme account for the observed spectral effects. By numerical solving the eqns. (4–5), we found a set of parameters { $K_b = 0.6~\mu\text{M}^{-1}$, $m = 0.3$, $K_{bd} = 0.2~\mu\text{M}^{-1}$, $m_d = 0.8$ } for **H1**, providing good agreement between the experimental and calculated values of A_{820}, A_{700} and A_{640} (the difference between the absorbances was less that 3%, 8% and 48%, respectively, at the maximum protein concentration). Figure 17A–C represents the measured and calculated absorbances at 640, 700 and 820 nm over the overall protein concentration range. This suggests that the behavior of dye **H1** in lysozyme fibrils system could be adequately described in terms of the dye monomer- and dimer-protein binding. On the contrary, for the set of parameters { $K_b = 0.8~\mu\text{M}^{-1}$, $m = 0.35$, $K_{bd} = 0.3~\mu\text{M}^{-1}$, $m_d = 2$ } recovered in analogous way for **H2**, the difference between the experimental and calculated values of A_{830}, A_{700} and A_{640} was ~9%, 21% and 66%, respectively, at the maximum protein concentration). Figure 17D–F represents the measured and calculated absorbances at 640, 700 and

Interactions between the Novel Cyanine Dyes ... 101

830 nm over the overall lysozyme concentration range. Notably, as seen in Figure 17D, theory predicts the linear drop in the **H2** H-aggregate concentration, the effect being opposite to that observed in the experiment.

Hypothesis 2. It seems likely that the observed minor discrepancies between the measured and calculated absorbances of **H1** H-aggregates and **H2** dimers result from the formation of J-aggregates by the part of the dye monomer species incorporated into fibrils. The protein-bound H-dimer species of **H2** could also transform into J-aggregates, because for the dye **H2** the H-dimer absorption is decreased upon increasing the protein concentration (Figure 17E).

Hypothesis 2a. Furthermore, significant difference between the measured and predicted intensities of **H2** H-band could result from the formation of H-aggregates with lower aggregation number compared to those observed in buffer. This could be, for instance, **H2** trimers. Thus, it appeared that, in contrast to **H1**, **H2** monomer binding to lysozyme fibrils is followed by the formation of the trimer species by the part of the protein-bound dye. This process is very complex and can be explained only qualitatively. Our results are in accordance with those obtained for the negatively charged pinacyanol dye associated with fibrillar Aβ [114], for the cyanine dye bound to the DNA groove [48], and for Thiazole Orange adsorbed on the calix[4]arene sulfonate template [134]. Specifically, the formation of H-aggregates was observed at increasing fibril or DNA concentration, although no aggregation occurred upon the dye binding to random coil and α-conformations. Summarizing the above results, we concluded that *Hypotheses 1a, 2* and *2a* can adequately reproduce the behavior of **H1** and **H2**, respectively, in the presence of lysozyme fibrils.

The simple docking studies were performed to assess the potential binding sites for heptamethine cyanine dyes. As can be seen from Figure 18, heptamethine dyes **H1** and **H2** interact with the deep cleft of native lysozyme lined with both hydrophobic and negatively charged residues. It appeared that the cleft volume is sufficient to accommodate the monomer and dimer dye species (Figure 18). The hydrophobic dye-protein interactions are likely to be predominant, similarly to the case of cyanine binding to HSA. Furthermore, in view of the fact that cyanine J-aggregates are formed on the negatively charged DNA templates [135], it can be assumed that electrostatic interactions between the positively charged dyes under study and negatively charged amino acids E53, E35, D52 and D70 of the lysozyme active site significantly stabilize the structure of J-aggregates of **H1** and **H2**.

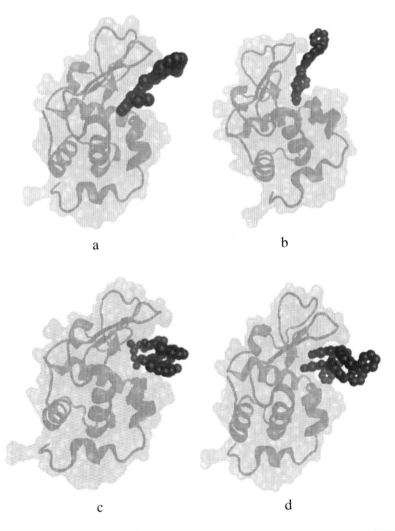

Figure 18. Schematic representation of the **H1/H2** monomer (A,B) and dimer (C,D) complexes with the native lysozyme, obtained using PatchDock/FireDock servers. The dyes are bound to a deep protein cleft, which comprises the lysozyme active site.

Due to their high affinities for lysozyme fibrils, the monomers of **H1** and **H2** are most likely incorporated into the fibril grooves representing specific binding sites for the amyloid markers [136]. Likewise, the binding to fibril grooves has been observed for mono-, trimethinecyanines and cyanine dye YOYO-1 [114, 134]. Notably, the calculated lengths, widths and heights of the optimized **H1** and **H2** structures were ~2.1, 1.3 and 0.6 nm while the distance between the every second residue and the interstrand distance in the β-sheet are ~0.7 nm and ~0.4 nm, respectively [137]. Therefore, the dye monomer species should associate with 5 β-strands in such a way that their short axis is perpendicular to the fibril axis. Our suggestions are also confirmed by the docking studies, which revealed the most energetically favourable **H1/H2**.binding in the Q75–N59/S60–W62,G54–L56 grooves of the lysozyme fibril core (Figure 19). Notably, aromatic substituents in the **H2** molecule, possessing higher affinity for lysozyme fibrils than **H1**, interact with tryptophan residues of the S60_W62/ L54_G56 channel, forming the most energetically favourable complex with the dye (Figure 19). Similarly, classical amyloid marker Thioflavin T preferentially interacted with the grooves, containing aromatic residues, as revealed from *in silico* studies [138].

In turn, according to our docking results, **H1** and **H2** dimer species seem to occupy the non-specific protein sites, lying perpendicular to the fibril axis. Similarly, high affinity of Thiazole Orange monomers for DNA was explained by the dye stacking with DNA bases, while dimers possessing low affinity (due to steric restrictions) were assumed to interact with external surface of DNA molecule in a way that prevents stacking of one monomer subunit with the DNA bases [51, 139]. Notably, the above estimate of the size of AK7-6 H-aggregates formed on the fibril surface ($n \approx 3$), was about one order of magnitude lower than that of cyanine aggregates obtained in liposomal and DNA systems, presumably due to steric restrictions for the dimer binding to fibril grooves [140].

According to the docking studies, unlike native lysozyme, fibril binding sites for the examined dyes are represented by the grooves formed by polar and nonpolar amino acid residues (Figure 19). This can explain a less pronounced formation of J-aggregates in the presence of fibrillar lysozyme compared to the native protein [140]. Notably, electrostatic attraction and hydrophobic interactions play significant role in the cyanine dye L-21 aggregate formation on DNA template [141]. H-aggregate dissociation into monomers and subsequent J-aggregate formation are presumably controlled by the strong monomer-protein hydrophobic and dye-dye electrostatic interactions [26]. Obviously, partial ejection of the bulky **H2** dye from the fibril groove (Figure 19) shifts the equilibrium towards the formation of H-aggregates on the fibril surface. Indeed, AK7-6 binding groove resides on the nonpolar surface of the lysozyme β-sheet [51]. Therefore, lateral association of the two β-sheets results in the formation of the dry "steric zipper" interface that is sterically less accessible for the dyes than the polar face of the fibril (Figure 19) [136].

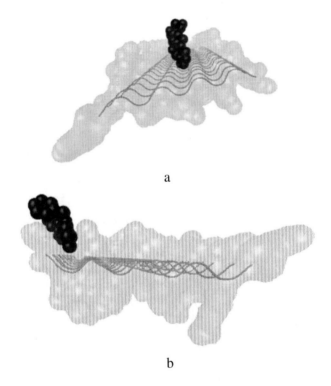

Figure 19. Schematic representation of the **H1** (A) and **H2** (B) complexes with the lysozyme fibril core (represented by the β-sheet formed from the K-peptide), obtained using PatchDock/FireDock servers.

Interestingly, spectral behavior of the **H1** and **H2** in the presence of fibrillar lysozyme was similar to that observed upon their binding to BSA, being indicative of the predominant role of the dye monomer-protein hydrophobic and dye-dye electrostatic interactions in the disruption of H-aggregates incorporated into the amyloid species, as well [57]. In view of the above rationales, we concluded that fibril grooves are not unique high-affinity binding sites for **H1** and **H2**, although analysis of the absorption spectra allows highly sensitive *in vitro* differentiation between the native and fibrillar lysozyme. Taken together, these results indicate that **H1** and **H2** can be employed for amyloid fibril detection using the absorption spectroscopy.

CONCLUSION

In conclusion, using the optical spectroscopy and molecular docking techniques, a series of monomethine, pentamethine and heptamethine cyanine dyes with the wide aromatic systems and different substitutions have been tested as potential probes for selective detection and characterization of biomacromolecules. The cyanine dyes under study, except of monomethines, undergo H-type self-association when free in a buffer solution. Based on the results of absorption studies, it was concluded that heptamethine and penthamethine dyes are capable of associating with nucleic acids, proteins and biomembranes and this process is accompanied by the changes in the dye aggregation behavior. More specifically, it was shown that DNA serves as a template for pentamethine (except **P5**) aggregate growth. The transfer of pentamethine dyes (except **P5**) into a lipid environment enchances their aggregation propensity. In turn, the association of heptamethines and penthamethine dye **P5** with the lipid vesicles led to the disruption of the dye aggregates rendering **H1**, **H2** and **P5** highly suitable as membrane probes. The fact that the reduction of lipid-induced heptamethine aggregation was more pronounced in the zwitterionic phosphatidylcholine bilayer in comparison with the negatively charged PG-containing liposomes, suggests that the disruption of highly organized molecular arrangements are most likely electrostatically controlled. The association of pentamethine and heptamethine dyes with bovine serum albumin was followed by the mutual conversions of the dye monomer/aggregate species, with the magnitude of the aggregate disruption being more pronounced for **H1** and **H2**. The possible explanation of the dissagregation of cyanine dyes under study involves the higher strength of the dye-BSA interactions compared to the coupling between the cyanine monomeric species in the aggregate. It has been shown that the heptamethine dye **H1** binds to the subdomain IIA, whereas the **H2** seems to occupy the subdomain IB. The quantitative characterization of the binding of heptamethine dyes to the native and fibrillar lysozyme was performed. A hypothetical model for the dye aggregation behavior was proposed, assuming the association of the dye monomers and dimers with the protein, in addition to the dye aggregation in a buffer. The obtained results showed that **H1** and **H2** display high affinity for lysozyme fibrils and have different aggregation extent in the presence of monomeric and fibrillar protein, rendering these probes suitable for noncovalent labeling of amyloid fibrils. The dye-fibril hydrophobic interactions and steric hindrances are supposed to have the predominant influence on the association of heptamethines with fibrillar lysozyme. The potential application of monomethine cyanine dyes as fluorescent probes for non-covalent labeling and characterization of dsDNA has been evaluated using the fluorescence spectroscopy technique. The examined dyes were found to have negligible fluorescence in the buffer solution, but exhibited a significant emission increase upon their binding to dsDNA. The binding studies indicate that monomethine cyanine dyes associate with the double stranded DNA presumably through the intercalating mode. The huge fluorescence enhancement observed upon the formation of the dye-DNA complexes allowed us to recommend the examined monomethine dyes for DNA detection.

REFERENCES

[1] Sahyun, M. R. V., Sharma, D. K. and Serpone, N. (1995). Mechanisms of spectral sensitization of silver-halides – role of sensitizing dye complexation. *J. Imaging Sci. Techn.*, 39: 377–385.

[2] Li, Q., Lin, G.-L., Peng, B.-X. and Li, Z.-X. (1998). Synthesis, characterization and photographic properties of some new styryl cyanine dyes. *Dyes and Pigments*, 38 (4) : 211–218.

[3] Lin, C. (1975) Polymethine IR laser dyes for passive mode-locking of ruby lasers. *Opt. Commun.*, 13(2) : 106–108.

[4] Langford, N., Smith, K., Sibbett, W. (1987) Passively mode-locked color-center laser *Opt. Lett.*, 12 (11): 903–905.

[5] Czerney, P., Graneb, G., Birckner, E., Vollmer, F., Rettig, W. (1995). Molecular engineering of cyanine-type fluorescent and laser dyes. *J. Photochem. Photobiol. A Chem.*, 89: 31–36.

[6] Mai, X., Hua, J., Wu, W., Jin, Y., Meng, F., Zhan, W., Tian, H. (2008). A high-efficicency cyanine dye for dye-sensitized solar cells. *Tetrahedron*, 64: 345–350.

[7] Wu, W., Hua, J., Jin, Y., Zhan, W., Tian, H. (2008). Protovoltaic properties of three new cyanine dyes for dye-sensitized solar cells. *Photochem. Photobiol. Sci.*, 7: 63–68.

[8] Chatterju, S., Gottschalk, P., Davis, P., Schuster, G. (1988). Electrontransfer reactions in cyanine-borate ion pairs: photopolymerization initiators sensitive to visible light. *J. Am. Chem. Soc.*, 110 (7): 2326–2328

[9] Ja, K., Zasada, M., Paczkowski, J. (2007). Photopolymerization reaction initiated by a visible light photoinitiating system: cyanine dye/ boratesalt/1,3,5,-triazine. *J. Polym. Sci. PartA: Polym. Chem.*, 45(16): 3626–3636.

[10] Makhlouf, M. T., Khalil, Z. H. Cyanine dyes as corrosion inhibitors. I. Electrochemical corrosion behaviour of copper metal treated with hydroxystyryl cyanine dye. *J. Chem. Technol. Biotechnol.*, 38(2): 89-97.

[11] Ning, J., Huang, B., Wei, Z., Li, W., Zheng, H., Ma, L., Xing, Z., Niu, H., Huang, W. (2017). Mitochondria targeting and near-infrared fluorescence imaging of a novel heptamethine cyanine anticancer agent. *Mol. Med. Rep.*, 15(6): 3761–3766.

[12] Si, S., Wu, J. B., Pan, D. (2016). Review on near-infrared heptamethine cyanine dyes as theranostic agents for tumor imaging, targeting and photodynamic therapy. *J. Biomed. Opt.*, 21(5): 050901.

[13] Abd El-Aal, R. M. Younis, M. (2004). Synthesis and antimicrobial activity of certain novel monomethine cyanine dyes. *Dyes Pigm.*, 6(3): 205–214.

[14] Sims, P. J., Waggoner, A. S., Wang, C. H., Hoffman J. F. (1974). Mechanism by which cyanine dyes measure membrane potential in red blood cells and phosphatidylcholine vesicles. *Biochemistry*, 13(16): 3315–3330.

[15] Armitage, B., O' Brien, D.F. (1992). Vectorial photoinduced electron transfer between phospholipid membrane-bound donors and acceptors. *J. Am. Chem. Soc.* 114: 7396–7403.

[16] Reers, M., Smith, T. W., Chen, L. B. (1991). J-aggregate formation of a carbocyanine as a quantitative fluorescent indicator of membrane potential. *Biochemistry*, 30(18): 4480–4486.

[17] Lou, Z., Li, P., Han K. (2015). Redox-responsive Fluorescent probes with different design strategies. *Acc. Chem. Res.*, 48(5): 1358–1368.

[18] Yarmoluk, S. M., Kostenko, A. M., Dubey, I. Y. (2000). Interaction of cyanine dyes with nucleic acids. Part 19: new method for the covalent labeling of oligonucleotides with pyrylium cyanine dyes. *Bioorg. Med. Che. Lett.*, 10(19): 2201–2204.

[19] Kubankova, M., Lopez-Duarte, I., Bull, J.A., Vadukul, D. M., Serpel L. C., Victor, M. S., Stride, E., Kuimova M. K. (2017). Probing supramolecular protein assembly attached fluorescent molecular rotors. *Biomaterials*, 139: 195–201.

[20] Waggoner, A. (1995). Covalent labeling of proteins and nucleic acids with fluorophores. *Methods. Enzymol.*, 246: 362–373.

[21] Patonay, G., Salon, J., Sowell, J., Strekowski, L. (2004). Noncovalent labeling of biomolecules with red and near-infrared dyes. *Molecules*, 9(3): 40-49.

[22] Tatikolov, A.S., Costa, S.M.B. (2004). Complexation of polymethine dyes with human serum albumin: a spectroscopic study. *Biophys. Chem.*, 107(1): 33–49.

[23] Patlolla, P. R., Desai, N., Gupta, S., Datta, B. (2019). Interaction of a dimeric carbocyanine dye aggregate with bovine serum albumin in non-aggregated and aggregated forms. *Spectrochim. Acta. A. Mol. Biomol. Spectrosc.*, 209: 256–263.

[24] Mahmood, T., Paul, A., Ladame, S. (2010). Synthesis, spectroscopic and DNA-binding properties of fluorogenic acridine-containing cyanine dyes. *J. Org. Chem.* 75(1): 204–207.

[25] Kaloyanova, S., Trusova, V.M., Gorbenko, G.P., Deligeorgiev T. (2011). Synthesis and fluorescence characteristics of novel asymmetric cyanine dyes for DNA detection. *J. Photochem. Photobiol. A.* 217: 147–156.

[26] Kurutos, A., Ryzhova, O., Trusova, V., Tarabara, U., Gorbenko, G., Gadjev, N., Deligeorgiev, T. (2016). Novel asymmetric monomethine cyanine dyes derived from sulfobetaine benzothiazolium moiety as potential fluorescent dyes for non-covalent labeling of DNA. *Dyes Pigm.*, 13: 122–128.

[27] Hilal, H., Taylor, J. A. (2007). Determination of the stoichiometry of DNA-dye interaction and application to the study of a bis-cyanine dye-DNA complex. *Dyes Pigm.*, 75(2): 483–490.

[28] Sidorowicz, A., Mora, C., Jablonka, S., Pola, A., Modrzycka, T., Mosiadz, D., Michalak, K. (2005). Spectral properties of two betaine-type cyanine dyes in syrfactant micelles and in the presence of phospholipids. *J. Mol. Struct.*, 744-747: 711–716.

[29] Krieg, M., Srichai, M.B., Redmond, R.W. (1993). Photophysical properties of 3,3'-dialkylthiacarbocyanine dyes in organized media: unilamellar liposomes and thin polymer films. *Biochim. Biophys. Acta Biomembr.*, 1151(2): 168–174.

[30] Volkova, K., Kovalska, V., Balanda, A., Losytskyy, M., Golub, A., Vermeij, R., Subramaniam, V., Tolmachev, O., Yarmoluk, S. (2008). Specific fluorescent detection of fibrillar alpha-synuclein using mono- and trimethine cyanine dyes. *Bioorg. Med. Chem.*, 16:1452–1459.

[31] Kovalska, V., Losytskyy, M., Tolmachev, O., Yu, S., Segers-Nolten, G., Subramaniam, V., Yarmoluk, S. (2012). Tri- and pentamethine cyanine dyes for fluorescence detection of α-synuclein oligomeric aggregates. *J. Fluor.*, 22(6):1441–1448.

[32] Daehne, S. (1978). Color and constitution: one hundred years of research. *Science*, 199 (4334): 1163–1167.

[33] Mooi, S., Keller, S.N., Heyne, B. (2014). Forcing aggregation of cyanine dyes with salts: a fine line between dimmers and higher ordered aggregates. *Langmuir*, 30:9654–9662.

[34] Kirstein, S., Daehne, S. (2006). J-aggregates of amphiphilic cyanine dyes: self-organization of artificial light harvesting complexes. *Int. J. Photoenergy*, 2006: 1–12.

[35] Czikkely, V., Forsterling, H. D., Kuhn, H. (1970). Light absorption and structure of aggregates of dye molecules. *Chem. Phys. Lett.*, 6(1):11–14.

[36] Prokhorov, V. V., Pozin, S.I., Lypenko, D. A., Perelygina, M. L., Maltsev, E. I., Vannikov, A. V. (2012). Molecular arrangements in polymorphous monolayer structures of carbocyanine dye J-aggregates. *Chem. Phys. Lett.*, 535:94–99.

[37] Kasha, M., Rawls, H. R., El-Bayoumi, M. A. (1965). The exciton model in molecular spectroscopy. *Pure. Appl. Chem.* 11:371–392.

[38] Oelkrug, D., Egelhaaf, H.-J., Gierschner, J., Tompert, A. (1996). Electronic deactivation in single chains, nano-aggregates and ultrathin films of conjugated oligomers. *Syn. Met.*, 76:249–253.

[39] Spano, F.C. (2006). Excitons in conjugated oligomer aggregates, films, and crystals. *Annu. Rev. Phys. Chem.*, 57:217–243.

[40] Tachibana, H., Sato, F., Terrettaz, S., Azumi, R., Nakamura, T., Sakai, H., Abe, M., Matsumoto, M. (1998) Light-induced J-aggregation in mixed Langmuir-Blodgett films of selenium-

containing cyanine and azobenzene. *Thin. Solid Films.*, 327-329: 813–815.

[41] Beckford, G., Owens, E., Henary, M., Patonay, G. (2012). The solvatochromic effects of side chain substitution on the binding interactions of novel tricarbocyanine dyes with human serum albimin. *Talanta*, 92: 45–52.

[42] Wang, M., Silva, G., Armitage, B. (2000). DNA-templated formation of a helical cyanine dye J-aggregate. *J. Am. Chem. Soc.*, 122:9977–9986.

[43] Berezin, M., Lee, H., Akers, W., Achilefu, S. (2007). Near infrared dyes as lifetime solvatochromic probes for micropolarity measurements of biological systems. *Biophys. J.*, 93:2892–2899.

[44] Behera, G.B., Behera, P.K., Mishira, B.K. (2007). Cyanine dyes: self-aggregation and behaviour in surfactants. A review. *J. Surf. Sci. Technol.*, 93:2892–2899.

[45] Peyratout, C., Donath, E., Daehne, L. (2002). Investigation of pseudoisocyanine aggregates formed on polystyrene sulfonate. *Photochem. Photobiol. Sci.*, 1: 87–91.

[46] Peyratout, C., Daehne, L. (2002). Aggregation of thiacyanine derivatives on polyelectrolytes. *Phys. Chem. Chem. Phys.* 4:3032–3039.

[47] Guralchuk, G. Yu., Katruniv, I. K., Grynyov, R. S., Sorokin, A. V., Yefimova, S. L., Borovoy, I. A., Malyukin, Yu. V. (2008). Anomalous surfactant-induced enhancement of luminescence quantum yield of cyanine dye J-aggregates. *J. Phys. Chem. C.*, 112:14762–14768.

[48] Hannah, K. S., Armitage, B. S. (2004). DNA-templated assembly of helical cyanine dye aggregates: a supramolecular chain polymerization. *Acc. Chem. Res.*, 37:845–853.

[49] Sovenyhazy, K. M., Bordelon, J. A., Petty, J. T. (2003). Spectroscopic studies of the multiple binding modes of a trimethine-bridged cyanine dye with DNA. *Nucleic Acids Res.*, 31(10):2561–2569.

[50] Zhang, Y., Xiang, J., Tang, Y., Xu, G., Yan, W. (2007). Chiral transformation of achiral J-aggregates of a cyanine dye template by human serum albumin. *Chem. Phys. Chem.*, 8:224–226.

[51] Vus, K., Tarabara, U., Kurutos, A., Ryzhova, O., Gorbenko, G., Trusova, V., Gadjev, N., Deligeorgiev, T. (2017). Aggregation behavior of novel heptamethine cyanine dyes upon their binding to native and fibrillar lysozyme. *Mol. Biosyst.*, 13:970–980.

[52] Schaberle, F. A., Kuz'min, V., Borissevitch, I. E. (2003). Spectroscopic studies of the interaction of bichromophoric cyanine dyes with DNA. Effect of ionic strength. *Biochim. Biophys. Acta.*, 1621:183–191.

[53] Patonay, G., Kim, J.S., Kodagahally, R., Strekowski, L. (2005). Spectroscopic study of a novel bis(heptamethine cyanine) dye and its interaction with human serum albumin. *Appl. Spectrosc.* 59(5):682–690.

[54] Kovalska, V., Losytskyy, M., Chernii, V., Volkova, K., Treyakova, I., Cherepanov, V., Yarmoluk, S., Volkov, S. (2012). Studies of anti-fibrillogenic activity of phthalocyanines of zirconium containing out-of-plane ligands. *Bioorg. Med. Chem.* 20(1):330–334.

[55] Vus, K., Girych, M., Trusova, V., Gorbenko, G., Kurutos, A., Vasilev, A., Gadjev, N., Deligeorgiev, T. (2019). Cyanine dyes derived inhibition of insulin fibrillization. *J. Mol. Liq.*, 276:541–552.

[56] Kurutos, A., Ryzhova, O., Trusova, V., Gorbenko, G., Gadjev, N., Deligeorgiev, T. (2016). Synthesis of meso-chloro-substituted pentamethine cyanine dyes containing benzothiazolyl/benzoselenazolyl chormophores Novel synthetic approach and studies on photophysical properties upon interaction with bio-objects. *J. Fluor.*, 26 (1):177–187.

[57] Kurutos, A., Ryzhova, O., Trusova, V., Gorbenko, G., Gadjev, N., Deligeorgiev, T. (2016). Novel synthetic approach to near-infrared heptamethine cyanine dyes and spectroscopic characterization in the presence of biological molecules. *J. Photochem. Photobiol. A: Chemistry.*, 328:87–96.

[58] Morozova-Roche, L. A., Zurdo, J., Spencer, A., Noppe, W., Receveur, V., Archer, D. B., Joniau, M., Dobson, C. M. (2000). Amyloid fibril formation and seeding by wild-type human lysozyme and its disease-related mutational variants. *J. Struct. Biol.*, 130:339–351.

[59] Petty, J. T., Bordelon, J. A., Robertson, M. E. (2000). Thermodynamic characterization of the association of cyanine dyes with DNA. *J. Phys. Chem. B.*, 104:7221–7227.

[60] Trusova, V. (2015). Modeling of amyloid fibril binding to the lipid bilayer. *East. Eur. J. Phys.* 2:51–58.

[61] Duhovny, D., Nussinov, R., Wolfson, H. J. (2002). Efficient unbound docking of rigid molecules. *Lect. Notes Comput. Sci. Eng.*, 2452:185–200.

[62] Kaloyanova, S., Crnolatac, I., Lesev, N., Piantanida, I., Deligeorgiev, T. (2012). Synthesis and study of nucleic acids interactions of novel monomethine cyanine dyes. *Dyes Pigm.*, 92:1184–1191.

[63] Yarmoluk, S., Lukashov, S., Ogul'chansky, T., Losytskyy, M., Kornyushyna, O. (2001). Interaction of cyanine dyes with nucleic acids. XXI. Arguments for half-intercalation model of interaction. *Biopolymers*, 62:219–227.

[64] Bricks, J. L., Kachkovskii, A. D., Slominski, Y. L., Gerasov, A. O., Popov, V. S. (2015). Molecular design of near infrared polymethine dyes: a review. *Dyes Pigm.*, 92:1184–1191.

[65] Spano, F. C. (2010). The spectral signature of Frenkel polarons in H- and J-aggregates. *Acc. Chem. Res.*, 43(3):429–439.

[66] Choudhury, S. D., Bhasikuttan, A. C., Pal, H., Mohanty, J. (2011). Surfactant-induced aggregation patterns of thiazole orange: a photophysical study. *Langmuir*, 27:12312–12321.

[67] Ghasemi, J., Ahmadi, Sh., Ahmad, A.I., Ghobas, S. (2008). Spectroscopic characterization of thiazole orange-3 DNA interaction. *Appl. Biochem. Biotechnol.*, 149: 9–22.

[68] Kim, J. S., Kodagahally, R., Strekowski, L., Patonay, G. (2005). A study of intramolecular H-complexes of novel bis(heptamethine cyanine) dyes. *Talanta*, 67:947–954.

[69] Balen, G., Martinet, C., Caron, G., Bouchard, G., Reist, M., Carrupt, P., Fruttero, R., Gasco, A., Testa, B. (2004). Liposome/water lipophilicity: methods, information content and pharmaceutical applications. *Med Res Rev.*, 3:299–324.

[70] Rye, H., Yue, S., Wemmer, D., Quesada, M., Haugland, P., Mathies, R., Glazer, A. (1988). Stable fluorescent complexes of double-stranded DNA with bis-intercalating asymmetric cyanine dyes: properties and applications. *Nucleic. Acid. Res.*, 20(11):2803–2812.

[71] Yan, X., Grace, W., Yoshida, T., Habbersett, R., Velappan, N., Jett, J., Keller, R., Marrone, B. (1999). Characteristics of different nucleic acid staining dyes for DNA fragment sizing by flow cytometry. *Anal. Chem.*, 71(24):5470–5480.

[72] Gurrieri, S., Wells, K., Johnson, I., Bustamante, C. (1997). Direct visualization of individual DNA molecules by fluorescence microscopy: characterization of the factors affecting signal/background and optimization of imaging conditions using YOYO. *Anal. Biochem.*, 249:44–53.

[73] Armitage, B., (2005). Cyanine dye-DNA interactions: intercalation, groove binding and aggregation. *Top. Curr. Chem.*, 253:55–76.

[74] Kricka, L. (2002). Stains, labels and detection strategies for nucleic acids assays. *Ann. Clin. Biochem.*, 39(2):114–129.

[75] Biver, T., Boggioni, A., Secco, F., Turriani, E., Venturini, S., Yarmoluk, S. (2007). Influence of cyanine dye structure on self-aggregation and interaction with nucleic acids: A kinetic approach to TO and BO binding. *Arch. Biochem. Biophys.*, 465:90–100.

[76] Davidson, Y., Gunn, B., Soper, S. (1996). Spectroscopic and binding properties of near-infrared tricarbocyanine dyes to double-stranded DNA. *Appl. Spectros.*, 50(2):211–221.

[77] Markova, L., Malinovskii, V., Patsenker, L., Haner, R. (2013). J- vs. H-type assembly: penthamethine cyanine (Cy5) as a near-IR chiroptical reporter. *Chem. Commun.*, 49:5298–5300.

[78] Sarwar, T., Rehman, S., Husain, M., Ishqi, H., Tabish, M. (2015). Interaction of coumarin with calf thymus DNA: Deciphering the mode of binding by in vitro studies. *Int. J. Biol. Macromol.*, 73:9–16.

[79] Mikelsons, L., Carra, C., Shaw, M., Schweitzer, C., Scaiano, J. (2005). Experimental and theoretical study of the interactions of single-stranded DNA homopolymers and a monomethine cyanine dyes: nature of specific binding. *Photochem. Photobiol. Sci.*, 4:798–802.

[80] Lerman, L. (1963). The structure of the DNA-Acridine complex. *Biochemistry*, 49:94–102.

[81] Zasedatelev, A., Gursky, G., Zimmer, C.H., Thrum, H. (1974). Binding of netropsin to DNA and synthetic polynucleotides. *Mol. Biol. Reports.*, 1(6):337–342.

[82] Yarmoluk, S., Lukashov, S., Losytskyy, M., Akerman, B., Kornyushyna, O. (2002). Interaction of cyanine dyes with nucleic acids. XXVI. Intercalation of the trimethine cyanine dye Cyan 2 into double-stranded DNA: study by spectral luminescence methods. *Spectrochim. Acta Part A.*, 58:3223–3232.

[83] Rye, H., Quesada, M., Peck, K., Mathies, R., Glazer, A. (1991). High-sensitivity two-color detection of double-stranded DNA with a confocal fluorescence gel scanner using ethidium homodimer and thiazole orange. *Nucleic Acid Res.*, 19(2):327–333.

[84] Nafisi, S., Saboury, A., Keramat, N., Neault, J-F., Tajmir-Riahi, H.-A. (2007). Stability and structural features of DNA intercalation with ethidium bromide, acridine orange and methylene blue. *J. Mol. Struct.*, 827:35–43.

[85] Gunther, K., Mertig, M., Seidel, R. (2010). Mechanical and structural properties of YOYO-1 complexed DNA. *Nucleic Acid Res.* 38(19):6526–6532.

[86] Caroff, A., Litzinger, E., Connor, R., Fishman, I., Armitage, B. (2002). Helical aggregation of cyanine dyes on DNA templates: effect of dye structure on formation of homo- and heteroaggregates. *Langmuir*, 18:6330–6337.

[87] M., Volkova, K., Kovalska, V., Makovenko, I., Slominskii, Yu., Tolmachev, O., Yarmoluk, S. (2005). Fluorescence properties of pentamethine cyanine dyes with cyclopentene and cyclohexene

group in presence of biological molecules. *J. Fluoresc.*, 15(6):849–857.

[88] Norden, B., Tjerneld, F. (1982). Structure of methylene blue-DNA complexes studied by linear and circular dichroism. *Biopolymers*, 21(9):1713–1734.

[89] Ogul'chansky, T., Yashcuk, V., Losytskyy, M., Kocheshev, I., Yarmoluk, S. (2000). Interaction of cyanine dyes with nucleic acids. XVII. Towards an aggregation of cyanine dyes in solutions as a factor facilitating nucleic acid detection. *Spectrochim. Acta Part A*, 56:805–814.

[90] Cevc, G. (1990). Membrane electrostatics. *Biochim. Biophys. Acta.*, 1031(3):311–382.

[91] Zhang, Y.P., Lewis, R., McElhaney, R.N. (1997). Calorimetric and spectroscopic studies of the thermotropic phase behavior of the n-saturated 1,2-diacylphosphatidylglycerols. *Biophys. J.*, 72:779–793.

[92] Rand, R.P., Parsegian, V.A. (1989). Hydration forces between phospholipid bilayers. *Biochim. Biophys. Acta.*, 988:351–376.

[93] Pan, J., Heberle, F., Tristram-nagle, S., Szymanski, M., Koepfinger, M., Katsaras, J., Kucerka, N. (2012). Molecular structures of fluid phase phosphatidylglycerol bilayers as determined by small angle neutron and X-ray scattering. *Biochim. Biophys. Acta.*, 1818:2135–2148.

[94] Duportail, G., Klymchenko, A., Mely, Y., Demchenko, A.P. (2002). On the coupling between surface charge and hydration in biomembranes: experiments with 3-hydroxyflavone probes. *J. Fluoresc.* 12(2):181–185.

[95] Klymchenko, A.S., Mely. Y., Demchenko, A.P., Duportail, G. (2004). Simultaneous probing of hydration and polarity of lipid bilayers with 3-hydroxyflavone fluorescent dyes. *Biochim. Biophys. Acta.*, 1665:6–19.

[96] Duportail, G., Klymchenko, A., Mely, Y., Demchenko, A.P. (2001). Neutral fluorescent probe with strong ratiometric response to surface charge of phospholipid membranes. *FEBS Letters.*, 508:196–200.

[97] Sparrman, T., Westlund, P. (2003). An NMR line shape and relaxation analysis of heavy water powder spectra of the L-alpha, L-beta and P-beta phases in the DPPC/water system. *Phys. Chem. Chem. Phys.*, 5: 2114–2121.

[98] MacDonald, P., Seelig, J. (1987). Calcium binding to mixed phosphatidylglycerol-phosphatidylcholine bilayers as studied by deuterium nuclear magnetic resonance. *Biochemistry*, 26:1231–1240.

[99] Scherer, P., Seelig, J. (1987). Structure and dynamics of the phosphatidylcholine and phosphatidylcholine headgroup in L-M fibroblasts as studied by deuterium nuclear magnetic resonance. *EMBO J.*, 6:2915–2922.

[100] Murzyn, K., Rog, T., Pasenkiewier-Gierula, M. (1989). Phosphatidylethanolamine-phosphatidylglycerol bilayer as a model of the inner bacterial membrane. *Biophys. J.*, 88:1091–1103.

[101] Tari, A., Huang, L. (1989). Structure and function relationship of phosphatidylglycerol in the stabilization of the phosphatidylethanolamine bilayer. *Biochemistry*, 28:7708–7712.

[102] Ishchenko, A. (1991) Structure and spectral-luminescent properties of polymethine dyes. *Russ. Chem. Rev.* 60:865–884.

[103] Kucherac, O., Oncul, S., Darwich, Z., Yushchenko, D., Arntz, Y., Didier, P., Mely, Y., Klymchenko, A. (2010). Switchable nile red-based probe for cholesterol and lipid order at the outer leaflet of biomembranes. *J. Am. Chem. Soc.*, 132(13):4907–4916.

[104] Ryzhova, O., Vus, K., Trusova, V., Kirilova, E., Kirilov, G., Gorbenko, G. (2016). Novel benzanthrone probes for membrane and protein studies. *Methods Appl. Fluoresc.*, 4:034007.

[105] Zhang, Y. Z., Yang, Q. F., Du, H. Y., Tang, Y. L., Xu, G. Z., Yan, W. P. (2008). Spectroscopic investigation on the interaction of a cyanine dye with serum albumins. *Chinese J. Chem.* 26:397–401.

[106] Tatikov, A. S., Costa, S. M. B. (2004). Complexation of polymethine dyes with human serum albumin: a spectroscopic study. *Biophys. Chem.*, 107:33–49.

[107] Alarcon, E., Aspee, A., Gonzalez-Bejar, M., Edwards, A. M., Lissi, E., Scaiano, J. C. (2010). Photobehavior of merocyanine 540 bound to human serum albumin. *Photochem. Photobiol. Sci.*, 9:861–869.

[108] Kim, J. S., Rodagahally, R., Strekowski, L., Patonay, G. (2005). A study of intramolecular H-complexes of novel bis(heptamethine cyanine) dyes. *Talanta*, 67:947–954.

[109] Bohme, U., Scheder, U. (2007). Effective charge of bovine serum albumin determined by electrophoresis NMR. *Chem. Phys. Lett.*, 434:342–345.

[110] Sudlow, G., Birkett, D. J., Wade, D. N. (1976). Further characterization of specific drug binding sites on human serum albumin. *Mol. Pharmacol.*, 12:1052–1061.

[111] Sowell, J., Mason, J. C., Strekovski, L., Patonay, G. (2001). Binding constant determination of drugs toward subdomain IIIA of human serum albumin by near-infrared dye-displacement capillary electrophoresis. *Electrophoresis*, 22:2512–2517.

[112] Uversky, V. N., Fink, A. L. (2004). Conformational constraints for amyloid fibrillation: the importance of being unfolded. *Biochim. Biophys. Acta*, 1698:131–153.

[113] Sabate, R., Estelrich, J. (2003). Pinacyanol as effective probe of fibrillar β-amyloid peptide: comparative study with Congo Red. *Biopolymers*, 72:455–463.

[114] Chegaev, K., Federico, A., Marini, E., Rolando, B., Fruttero, R., Morbin, M., Rossi, G., Fugnanesi, V., Bastone, A., Salmona, M., Badiola, N. B., Gasparini, L., Cocco, S., Ripoli, C., Grassi, C., Gasco, A. (2015). NO-donor thiacarbocyanines as multifunctional agents for Alzheimer's disease. *Bioorg. Med. Chem.*, 23:4688–4698.

[115] Volkova, K. D., Kovalska, V. B., Inshin, D., Slominskii, Y. L., Tolmachev, O. I., Yarmoluk, S. M. (2011). Novel fluorescent trimethine cyanine dye 7519 for amyloid fibril inhibition assay. *Biotech. Histochem.*, 86:188–191.

[116] Yang, W., Wong, Y., Ng, O. T. W., Bai, B. L. P., Kwong, D. W. J., Ke, Y., Jiang, Z. H., Li, H. W., Yung, K. L. L., Wong, M. S. (2012). Inhibition of beta-amyloid peptide aggregation by mujltifunctional

carbazole-based fluorophores. *Angew. Chem. Int. Ed.*, 51:1804–1810.

[117] Kuret, J., Chirita, C. N., Congdon, E. E., Kannanayakal, T., Li, G., Necula, M., Yin, H., Zhohg, Q. (2005). Pathways of tau fibrillization. *Biochim. Biophys. Acta*, 1739:167–178.

[118] Vieira Ferreira, L. F., Oliveira, A. S., Wilkinsonb, F., Worrallb, D. (1996). Photophysics of cyanine dyes on surfaces. A new emission from aggregates of 2,2'-cyanines adsorbed onto microcrystalline cellulose. *J. Chem. Soc. Faraday Transaction*, 92:1217–1225.

[119] Kim, O. K., Je, J., Jernigan, G., Buckley, L., Whitten, D. (2006). Super-helix formation induced by cyanine J-aggregates onto random-coil carboxymethyl amylose as template. *J. Am. Chem. Soc.*, 128:510–516.

[120] Khairutdinov, R. F., Serpone, N. (1997). Photophysics of cyanine dyes: subnanosecond relaxation dynamics in monomers, dimers and H-and J-aggregates in solution. *J. Phys. Chem. B.*, 101:2602–2610.

[121] Garoff, R. A., Litzinger, E. A., Connor, R. E., Fishman, I., Armitage, B. A. (2002). Helical aggregation of cyanine dyes on DNA templates: effect of dye structutre on formation of homo-and heteroaggregates. *Langmuir*, 18:6330–6337.

[122] Tai, S., Hayashi, M. (1991). Strong aggregation properties of novel naphthalocyanines. *J. Chem. Soc.*, 2:1275–1279.

[123] Huang, W., Wang, L. Y., Fu, Y. L., Liu, J. Q., Tao, Y. N., Fan, F. L., Zhai, G. H., Wen, Z. Y. (2009). Study on thermodynamics of three kinds of benzindocarbocyanine dyes in aqueous methanol solution. *Bull. Korean Chem. Soc.*, 30:556–560.

[124] Schutte, W. J., Sluyters-Rehbach, M., Sluyters, J. H. (1993). Aggregation of an octasubstituted phthalocyanine in dodecane solution. *J. Phys. Chem.*, 97:6069–6073.

[125] Chang, E., Congdon, E. E., Honson, N. S., Duff, K. E., Kuret, J. (2009). Structure-activity relationship of cyanine tau aggregation inhibitors. *J. Med. Chem.*, 52:3539–3547.

[126] Berlepsch, H., Brandenburg, E., Koksch, B., Bottcher, C. (2010). Peptide adsorption to cyanine dye aggregates revealed by cryotransmission electron microscopy. *Langmuir*, 26:11452–11460.

[127] Kovalska, V. B., Losytskyy, M. Yu., Chernii, S. V., Chernii, V. Ya., Tretyakova, I. M., Yarmoluk, S. M., Volkov, S. V. (2013). Towards the anti-fibrillogenic activity of phtalocyanines with out-of-plane ligands: correlation with self-association proneness. *Biopolymers and Cell*, 29: 473–479.

[128] Sulatskaya, A. I., Kuznetsova, I. M., Turoverov, K. K. (2011). Interaction of thioflavin T with amyloid fibrils: stoichiometry, and affinity of dye binding, absorption spectra of bound dye. *J. Phys. Chem. B*, 115:11519–11524.

[129] Klunk, W. E., Pettegrew, J. W., Abraham, D. J. (1989). Quantitative evaluation of Congo red binding to amyloid-like proteins with a beta-pleated sheet conformation. *J. Histochem. Cytochem.*, 37: 1273–1281.

[130] Pisoni, D. S., Todeschini, L., Borges, A. C. A., Petzhold, C. L., Rodembusch, F. S., Campo, L. F. (2014). Symmetrical and asymmetrical cyanine dyes. Synthesis, spectral properties and BSA association study. *J. Org. Chem.*, 79:5511–5520.

[131] Seifert, J. L., Connor, R. E., Kushon, S. A., Armitage, B. A. (1999). Spontaneous assembly of helical cyanine dye aggregates on DNA nanotemplates. *J. Am. Chem. Soc.*, 121:2987–2995.

[132] Volkova, K. D., Kovalska, V. B., Losytskyy, M. Yu., Fal, K. O., Derevyanko, N. O., Slominskii, Yu. L., Tolmachov, O. I., Yarmoluk, S. M. (2011). Hydroxy and methoxy substituted thiacarbocyanines for fluorescent detection of amyloid formation. *J. Fluoresc.*, 21: 775–784.

[133] Lindberg, D. J., Esbjorner, E. K. (2016). Detection of amyloid-β fibrils using the DNA intercalating dye YOYO-1: Binding mode and fibril formation kinetics. *Biochem. Biophys. Res.*, 469:313–318.

[134] Lau, V., Heyne, B. (2010). Calix[4]arene sulfonate as a template for forming fluorescent thiazoleorange H-aggregates. *Chem. Commun.*, 46:3595–3597.

[135] Godjayev, N. M., Akyuz, S., Ismailova, L. (1998). The conformational properties of Glu 35 and Asp 52 of lysozyme active center. *Int. J. Phys. Engineer. Sci.*, 51:56–60.

[136] Vus, K., Trusova, V., Gorbenko, G., Sood, R., Kirilova, E., Kirilov, G., Kalnina, I., Kinnunen, P. (2014). Fluorescence investigation of interactions between novel benzanthrone dyes and lysozyme amyloid fibrils. *J. Fluoresc.*, 24:493–504.

[137] Krebs, M. R., Bromley, E. H., Donald, A. M. (2005). The binding of thioflavin-T to amyloid fibrils: localization and implication. *J. Struct. Biol.*, 149:30–37.

[138] Biancalana, M., Koide, S. (2010). Molecualar mechanism of Thioflavin-T binding to amyloid fibrils. *Biochim. Biophys. Acta*, 1804:1405–1412.

[139] Kuperman, M. V., Chernii, S. V., Losytskyy, M. Yu., Kryvorotenko, D. V., Derevyanko, N. O., Slominskii, Yu. L., Kovalska, V.B., Yarmoluk, S.M. (2015). Trimethine cyanine dyes as fluorescent probes for amyloid fibrils: the effect of N, N'-substituents. *Anal. Biochem.*, 484: 9–17.

[140] Ogulchansky, T. Y., Losytskyy, M. Yu., Kovalska, V. B., Lukashov S. S., Yaschuk, V. M., Yarmoluk, S. M. (2001). Interaction of cyanine dyes with nucleic acids. XVIII. Formation of the carbocyanine dye J-aggregates in nucleic acids grooves. *Spectrochim. Acta Part A*, 57:2705–2715.

[141] Guralchuk, G. Y., Sorokin, A. V., Katrunov, I. K., Yefimova, S. L., Lebedenko, A. N., Malyukin, Y. V., Yarmoluk, S. M. (2007). Specificity of cyanine dye L-21 aggregation in solutions with nucleic acids. *J. Fluoresc.* 17:370–376.

In: Cyanine Dyes
Editor: Douglas Zimmerman

ISBN: 978-1-53616-239-4
© 2019 Nova Science Publishers, Inc.

Chapter 3

CYANINE DYES, J- AND H- AGGREGATION IN THE PRESENCE OF NANOPARTICLES: EXPERIMENTAL AND THEORETICAL APPROACH AND APPLICATION

*Dragana Vasić-Aničijević, Tamara Lazarević-Pašti and Vesna Vasić**

Vinča Institute of Nuclear Sciences, University of Belgrade,
Belgrade, Serbia

ABSTRACT

The cyanine dyes belong to the group of polymethine synthetic organic compounds which have the application in a variety of spectroscopy detection techniques in many fields of science and technology (to increase the sensitivity range of photographic emulsions, as photosensitizers in photonic devices, in biological and medicinal images to label proteins, antibodies, peptides, nucleic acid probes and any kind of other biomolecules). They undergo to the spontaneous self-

---

* Corresponding Author's E-mail: evasic@vin.bg.ac.rs.

organization in solution forming J- and H-aggregates, with a characteristic change of their optical properties. This self-organization of cyanine dyes is usually supported by the presence of metal ions, macromolecules or nanoparticles. The organization of dye molecules in a parallel way (plane-to-plane stacking) leads a sandwich-type arrangement (H-aggregates) with a blue-shifted absorption since a head-to-tail arrangement (end-to-end stacking) induces J-aggregates formation with a red-shifted absorption band in the absorption spectrum with respect to the monomer absorption. There is also a dramatic change in their fluorescence properties. In this paper, we give an overview of various experimental and theoretical methods for the study of cyanine dyes self-organization, as well as their application.

Keywords: thiacyanine dyes, J- and H-aggregation, nanospectroscopy methods, DFT calculations

1. INTRODUCTION

The natural light-harvesting complexes (LHCs) of plants and photosynthetic bacteria are one of the most fascinating functional molecular assemblies (Jang and Mennucci 2018). Their main purpose is strong absorption of light followed by fast energy transfer to neighbouring LHCs and the photochemical reaction centre where an electron transfer process leads to charge separation. The effectiveness of these processes relies on two fundamental physicochemical principles. First *self-organization* of dye molecules, preferentially of chlorophyll derivatives mediated by proteins, into precisely ordered structures of large spatial size leads to obtain extraordinarily high cross-sections for light absorption (Oba and Tamiaki 1998). Second, extremely fast *energy migration* of the absorbed light within the LHC can ensure that the excitation energy is extremely fast available at any place where it is needed for photo-induced energy or electron transfer processes (Clark, Krueger, and Vanden Bout 2014).

The self-association of dyes in solution or at the solid-liquid interface (Behera, Behera, and Mishra 2007) is a frequently encountered phenomenon in dye chemistry owing to strong intermolecular van der Waals-like attractive forces between the molecules. The aggregates in solution exhibit distinct changes in the absorption band as compared to the monomeric species. In order to achieve a high degree of molecular order, it was tried to build artificial light-harvesting systems by adsorbing dye molecules on solid substrates. In such heterogeneous systems, the underlying crystal lattice of the solid substrate acts as a matrix for stable fixation and precise stacking of the dye molecules by epitaxial growth (Tomioka, Kinoshita, and Fujimoto 2007, Dahne 1995).

Cyanine dyes is the non-systematic name of a synthetic dye family belonging to the polymethine group of compounds, synthesized to build up artificial light-harvesting systems (Behera, Behera, and Mishra 2007, Shindy 2017). They consist of two nitrogen centres, one of which is positively charged and is linked by a conjugated chain of an odd number of carbon atoms to the other nitrogen. In general, there are three types of cyanine dyes classified on the basis of the charge of streptomethine unit: strepto cyanines or open chain cyanines (R_2N^+=CH[CH=CH]$_n$-NR$_2$); hemicyanines (Aryl=N$^+$=CH[CH=CH]$_n$-NR$_2$) and closed chain cyanines (Aryl=N$^+$=CH[CH=CH]$_n$-N=Aryl)(Pisoni et al. 2014). In these compounds, two quaternary nitrogens are joined by a polymethine chain, each of them is the independent part of a heteroaromatic moiety, such as pyrrole, imidazole, thiazole, pyridine, quinoline, indole, benzothiazole. The general structural formula of the cationic thiacyanine dyes is presented in Figure 1. These dyes have a wide application in many fields of science and practice. Firstly, they were used to increase the sensitivity range of photographic emulsions and also in many photonic devices, but later their application was extended to the medicinal and biosensing field, to label proteins, antibodies, peptides, nucleic acid probes, and any kind of other biomolecules which are used in a variety of fluorescence detection (Bricks et al. 2017, Benson and Kues 1977).

The combination of emissive labels such as fluorescent dyes and metal NPs to prepare inorganic-organic hybrid materials is a strategy to enhance their photophysical properties, useful for different applications, such as bioimaging, sensing, and optoelectronics (Kneipp et al. 2009, Liu, Sanyasi Rao, and Nunzi 2011, Demchenko 2013, Zhang, Wu, and Berezin 2015).

Such hybrid systems can exhibit properties that take advantage of both, dye and NPs, resulting in a material with intermediate properties or unique characteristics distinct from either the NPs or dye properties alone (Lebedev and Medvedev 2013a). While plasmonic coupling between NPs is known to be responsible for the enhancement of the local electrical field that determines the optical or spectroscopic properties, understanding of the interparticle molecular interactions and reactivities is still a challenge.

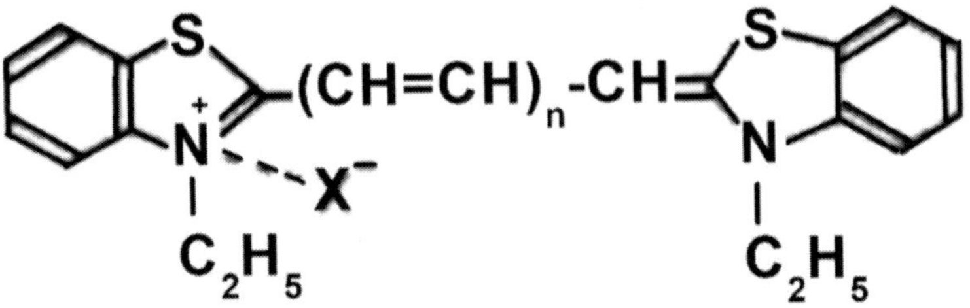

Figure 1. General structure formula of polymethine thiacyanine dyes: n - the number of methine groups, $X^- = Br^-$, Cl^-, ClO_4^-, and various anions.

A lot of efforts for the fabrication and characterization of new NPs – dye structures have been made, and existing nanospectroscopy tools enable the study of these materials and address some questions in this field, which will be useful for better understanding of such hybrid effects as well as the exploration of novel functionalities useful for the application to molecular plasmonic devices.

2. SELF-ORGANIZATION OF CYANINE DYES

The cyanine dyes possess very high ground state polarizability of the π-electrons along with the polymethine group in the ground state, which gives rise to strong dispersion forces (van der Waals forces) between two cyanine molecules in solution (Kirstein and Daehne 2006). One of the main features of cyanine dyes is their ability for spontaneous self-organization into highly ordered aggregates of various structures and morphologies.

Figure 2. Schematic presentation of the possible arrangements of cyanine dyes on the solid surface and in solution: Ladder arrangement (left), staircase arrangement (middle), brickwork arrangement (right).

As presented schematically in Figure 2, they can be organized in a parallel way (plane-to plane stacking) to form a sandwich-type arrangement (H-dimmer, hypsochromically shifted H-bands, H for hypsochromic) or in a head-to-tail arrangement (end-to end stacking) to form J-dimmer (bathrochromically-shifted excitonic absorption, J-bands) (Shindy 2017). This organization enables the high polarizability of the π-electrons along with the polymethine group in the ground state and the efficient exciton coupling including the fast exciton energy migration over thousands of molecules within a few picoseconds. Extensive studies on J- and H- aggregates have resulted in the proposal that these aggregates exist as a one - dimensional assembly in a solution that could be in brickwork, ladder, or staircase type of arrangement (Prokhorov et al. 2012).

2.1. H- and J-Aggregation of Cyanine Dyes

The most widely known aggregates of polymethine dyes are dimmers, J-, H- and H*-aggregates (Mishra et al. 2000, Gadde, Batchelor, and Kaifer 2009, Gadde et al. 2008). As an example, the structures of some of the most frequently studied and used cyanine dyes, which undergo to self-organization and aggregation, are presented in Table 1, but the number of new synthesized and studied dyes is continually increasing.

Table 1. Structures of some of the most frequently studied cyanine dyes

Dye structure	Chemical formula	Trivial name
	1,1′-Diethyl-2,2′-cyanine chloride	Pseudoisocyanine PIC
	3,3-disulfopropyl-5,5-dichloro -9-methyl thiacarbo cyanine	TCC
	Indocyanine green	ICD
	3,3'-disulfopropyl-5,5'-dichloro-thiacyanine sodium salt	Thiacyanine dye TC

In this paper, we focused mainly on the study of J-aggregation of the anionic thyacyanine dye (3,3'-disulfopropyl-5,5'-dichlorothiacyanine sodium salt, TC) under the various conditions (in the presence of mono and divalent metal salts and different types of nanoparticles (NPs).

The selected dye has the integrated optical (spectrophotometrical and fluorescence) properties, which strongly depend on the concentration of metal cations, which promote its J-aggregation (Vujačić et al. 2012, Laban, Vodnik, and Vasić 2015, Chibisov, Slavnova, and Görner 2008). This dye is chosen because of its appropriate solubility in water, high extinction coefficient as well as because of it the fluorescence properties. Moreover, it easily undergoes to self-organization, forming J-aggregates in the presence of metal ions as well as NPs.

The information about the aggregation of polymethine dyes can be most simply obtained from the shape of the optical absorption bands (Bricks et al. 2017, Egorov 2017). This information enables to draw the conclusion about the entire structure of the aggregate and the number of molecules in the optical chromophore. For example, the optical chromophore of J-aggregates, which consists of four monomers forming a brickwork-type structure, produces aggregates in the form of long thin rods.

Jelly and Scheibe (Egorov and Alfimov 2007, Egorov 2009, 2017) were the first who described the formation of aggregates of pseudo PIC (Table 1), which show a bathrochromically-shifted excitonic absorption (J-bands) that can migrate over thousands of molecules. The formation of such aggregates, called J-aggregates after Jelly, but also Scheibe-aggregates is a characteristic of cyanine dyes, but it is also shown by the homo-associates of several types of compounds. The study of J-aggregates is also of theoretical interest because of the role of excitons in biological light-harvesting systems (Kirstein and Daehne 2006). The spectral characteristics of J-bands and hypsochromically shifted H-bands (H for hypsochromic) of the aggregates have been explained in terms of molecular exciton coupling theory, i.e., the coupling of transition moments of the constituent dye molecules (Mishra et al. 2000).

In general, the absorption bands of the dimmers and H - aggregates are blue-shifted relative to the monomer band, since the absorption band of J-aggregates is red-shifted. The bandwidth and intensity also change due to both H- and J – aggregates formation. This unique feature of dye aggregates is widely used in modern high technologies (Egorov 2017, Bricks et al. 2017).

H- and J-aggregation in solution is strongly dependent on dye concentration, ionic strength, pH, dielectric constant, the presence of metal ions, NPs, polymers, macromolecules and also on covering of solid surface. However, the molecules show novel properties, such as linear and nonlinear optical response, photoelectric, photorefractive, photochromism, superradiance, superfluorescence, electroluminescence, photoluminescence, an attenuated total reflection which have the applications in major fields of science (Struganova et al. 2003, Struganova, Lim, and Morgan 2002, Tillmann and Samha 2004, Yao, Isohashi, and Kimura 2007).

2.2. Optical Properties of Cyanine Dyes

The absorption spectra of cyanine dyes usually exert very narrow J-bands (only 10–20 nm in width) with strongly increased molar extinction coefficient in comparison with isolated monomers and the dramatic red shift (often by 100 nm) due to the formation of J-aggregates. Very narrow light absorption and emission bands of these molecular associates suggest a high level of their ordering and their spontaneous and reversible self-assembly (Avakyan, Shapiro, and Alfimov 2014, Benson and Kues 1977). The fluorescence properties of J-aggregates match the absorption spectra and are of equally narrow width (Egorov 2017).

In contrast, in some conditions, the dyes can assemble into quite different H-aggregates which have the absorption band shifted hypsochromically but the correspondent fluorescence spectra (which can be very low in intensity) are shifted to longer wavelengths. The division of cyanine dyes self-organization into H- and J- aggregates is a simplification, since some intermediate cases may be expected so that H- and J-bands or several J-bands can be observed simultaneously. The aggregates also demonstrate high optical anisotropy and highly polarized fluorescence in solutions or in oriented polymer films that can be observed in the microscope (Zhang, Wu, and Berezin 2015).

The molecular arrangement of cyanine dyes is often described as the one-dimensional supramolecular self-organization which proceeds *via* π–π dispersive interactions between highly polarizable groups of atoms together with electrostatic interactions between opposite charges. However, the observed spectroscopic changes are the result of excitonic coupling between individual molecules forming the aggregate. By definition, excitons are the mobile neutral quasi-particles in the form of electron–hole pairs that appear in solid bodies and molecular associates upon electronic excitation (Davydov 1963, Egorov 2009).

The resonant excitation transfer interaction between constituting molecules is much stronger compared to the interaction with the environment because of the coherent exciton motion along the molecular chain. These electron-hole pairs may diffuse through the whole aggregate, while at each instant of time electron and hole occupy the same molecule and can be delocalized across many chromophores (Jang and Mennucci 2018).

H- and J- aggregates represent the limiting cases of specific molecular packings for a description of optical properties by theory of molecular excitons (Egorov and Alfimov 2007, Higgins, Reid, and Barbara 1996, Jang and Mennucci 2018). The dye molecule is regarded as a point dipole, and the excitonic state of the dye aggregate splits into two levels through the interaction of transition dipoles (Kasha 1963). Figure 3 illustrates the exciton model for the variation of energy on electronic transitions of molecular dipoles (Kasha 1963, Bricks et al. 2017). The H-type dimers formed by side-to-side association (sandwich structure) demonstrate strongly increased energy separation between absorbing and emitting states and, due to forbidden character, their fluorescence is low. An electron transition to the upper state in the parallel aggregates having parallel transition moments leads to hypsochromic (blue) shifts of the absorption band with respect to the monomer band. In J- aggregates, the transitions dipoles are in line with the molecular axis of the dimer and the transition to the lower excited level is allowed. Consequently, the lower state in a head-to-tail arrangement with perpendicular transition moments and bathochromic (red), leads to red-shifted absorption maximum relative to the absorption of monomer (Behera, Behera, and Mishra 2007). The anionic TCC monomer (Table 1), which contains one methine group coupled with two N, S containing thiazole rings shows a considerable planar geometry between two benzothiazole end groups and forms both H- and J- aggregates. J-aggregates exhibit the red-shifted narrow band in the absorption spectrum, while H-aggregates consist of a blue-shifted band with respect to the monomer absorption spectra (Yao et al. 2005). Their formation is strongly dependent on the dye concentration in solution, as well as the presence of various salts. At the dye concentration of 1×10^{-3} mM, it forms pure mesoscopic *H* aggregates. With an increase in the dye concentration up to 0.5 mM, an increased amount of mesoscopic *J* aggregates is observed at the expense of the *H* aggregates, and finally, in the presence of 10 mM TCC in solution, the dye forms the *J* aggregates predominantly (Yao et al. 2005).

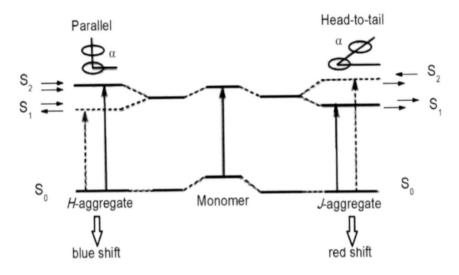

Figure 3. Schematic representation of the relationship between chromophore arrangement and spectral shift.

These distinct changes in the absorption spectra compared to the monomeric species have been explained in terms of molecular exciton coupling theory, i.e., the coupling of transition moments of the constituent dye molecules (Hestand and Spano 2018). Besides, the *mesoscopic behaviour* of the dye aggregation can be controlled by other external perturbations, e.g., additions of various electrolytes or macromolecules into the solution. They promote H- or J- aggregation by favoring a face-to-face or a twisted conformation.

3. J- AGGREGATION OF THIACYANINE DYES ON NPS

3.1. Noble Metal Nanoparticles

NPs, in general, have unique chemical and physical properties and the great number of functional platforms that can be utilized in many applications, as electrochemistry, photochemistry, chemical sensing and biomedicine (Alaqad and Saleh 2016). Noble metal NPs distinguish from other nanoplatforms like semiconductor quantum dots, magnetic nanoparticles and polymeric nanoparticle by their single surface plasmon resonance (SPR), which has a small particle size, enhances all the radiative and irradiative properties of the nanoparticles (Link and El-Sayed 1999).

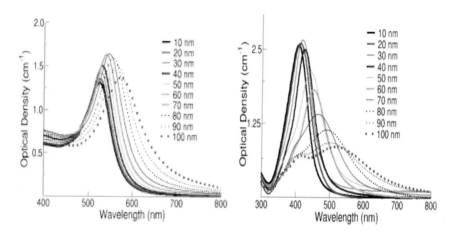

Figure 4. Dependence of extinction (scattering + absorption) spectra of gold (left) and silver (right) nanoparticles on diameters (http://www.sigmaaldrich.com/materials-science/nanomaterials/silver-nanoparticles.html).

Among various nanostructures, noble metal NPs consisting of Au or Ag have been studied extensively because of their unique optical properties arising from localized surface plasmon resonance in the visible region (Alaqad and Saleh 2016, Barooah et al. 2011, Daniel and Astruc 2003). The spectral response of spherical AuNPs and AgNPs in solution as a function of particle diameter is shown in Figure 4. As the diameter increases, the SPR band shifts to longer wavelengths and broadens. AuNPs are widely used in biotechnology and biomedical field due to their inert nature, stability, high disparity, non-cytotoxicity, and biocompatibility (Alaqad and Saleh 2016, Daniel and Astruc 2003, Devaraj et al. 2013, Schwartzberg et al. 2004). They are proper for the surface immobilization, and the modification of their surface leads to the enhancement of the interaction of these NPs with biological cells. They also have many applications, such as catalysis, optical molecular sensing, cancer therapeutics, and construction blocks in nanotechnology (Saleh 2014).

AgNPs are used in antimicrobial applications since the antimicrobial effect of Ag ions is well known (Tran, Nguyen, and Le 2013, Zhang et al. 2016, Yoshida, Kometani, and Yonezawa 2008). They also have special chemical and physical properties, such as surface-enhanced Raman scattering and optical behaviour, electrical conductivity, high thermal, chemical stability, nonlinear, and catalytic activity. Furthermore, AgNPs have individual plasmon optical spectra properties which allow them to be used in biosensing application.

However, there is a considerable interest in these materials, because of their ability to promote the J-aggregation of cyanine dyes by forming the hybrid systems with unusual optical properties. It is believed that the dye' ability to aggregate on the surface of colloidal metal NPs relies primarily on electrostatic attractions between the ionic molecular building blocks of the aggregates and charged colloids (Zhang, Fang, and Shao 2006, Yoshida, Yonezawa, and Kometani 2009, Vujačić, Vodnik, et al. 2013, Vujačić, Vasić, et al. 2013, Laban, Vodnik, and Vasić 2015, Laban et al. 2016, Laban et al. 2014). Due to the electric charge of colloids, the concentration of the dye near the surface is several orders of magnitude higher. Furthermore, the lack of covalent bonding between the dye molecules and the metal surfaces allows the molecules to be mobile enough on the colloid surface to allow for repositioning and self-assembly of the "brickwork" like pattern that gives rise to intermolecular electronic coupling necessary for the formation of the J-aggregate exciton (Jang and Mennucci 2018).

Cyanine dyes itself are usually positively or negatively charged due to the two corresponding end groups. However, they can be electrostatically attracted to the positively or negatively charged colloid surface through the binding via the charged end groups. Figure 5 illustrates the interaction between TC dye anion with the charged surface of AgNPs or AuNPs (Vujačić et al. 2012, Vujačić, Vodnik, et al. 2013, Laban et al. 2016, Laban et al. 2013).

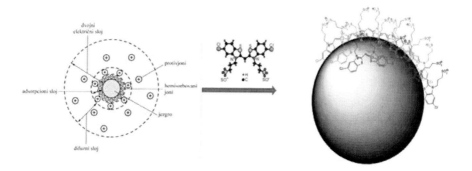

Figure 5. The illustration of alignment of optimized TC molecule structure on the NPs surface.

It is worthy to notice that TC can be adsorbed on the surface of AuNPs and AgNPs of various size and surface capping, but only some specific NPs induce the J-aggregation of cyanine dyes on their surface.

3.2. NPs Characterization

Various techniques are used for determining NPs properties such as the size, geometry, shape, crystallinity and surface area, in order to evaluate their nature (Zhang et al. 2016, Alaqad and Saleh 2016). These techniques are UV–Vis spectroscopy, transmission electron and scanning microscopy (TEM, SEM), X-ray photoelectron spectroscopy (XPS), atomic force microscopy (AFM), Raman spectroscopy and Fourier transform infrared spectroscopy (FTIR). The NPs morphology, shape and particle size could be specified by TEM, AFM and SEM. Moreover, particle size distribution (PSD) in colloid suspensions, including their hydrodynamic radius, can be by evaluated by dynamic light scattering (DLS). Furthermore, the determination of crystallinity is performed by X-ray diffraction (XRD), while UV–Vis spectroscopy technique is used to confirm the formation by showing the plasmon resonance (Vodnik and Nedeljković 2000).

3.2.1. Nanospectroscopy Techniques

UV-vis spectroscopy is a very useful and reliable technique for the primary characterization of synthesized NPs because it is fast, easy, simple, sensitive and selective. Noble metal NPs have unique optical properties which make them strongly interact with specific wavelengths of light. In AgNPs and AuNPs the conduction and valence bands lie very closly to each other, and the electrons, which move freely, give rise to a surface plasmon resonance (SPR) absorption band, occurring due to the collective oscillation of electrons in resonance with the light wave. Observation of the peak assigned to a surface plasmon is well documented for various metal nanoparticles with sizes ranging from 2 to 100 nm (Zhang et al. 2016, Alaqad and Saleh 2016).

FTIR analysis is usually used for the characterization of the functional groups observed on the nanomaterials and NPs surfacees due to their synthesis or functionalization. This method, which is able to provide accuracy, reproducibility, and also a favourable signal-to-noise ratio is a non-invasive technique with the rapid data collection, strong signal, large signal-to-noise ratio and less sample heat-up. Therefore, FTIR is a suitable, valuable, non-invasive, cost-effective, and simple technique to identify the role of various molecules in the reduction of noble metals and NPs functionalization (Golla, Koduru, and Borelli 2011, Devaraj et al. 2013, Momić et al. 2016, Lazarević Pašti et al. 2018).

The Raman technique has been explored on the characterization of AgNPs and AuNPs, and also on other materials (Neves and Andrade 2015, Mirković et al. 2016). The Raman signal of colloids and core shells shows the SPR and SERS enhancement. SERS is a useful tool for understanding the dye-NPs interaction, as well as detection of dye-containing NPs systems. Both LSPR absorption and SERS were found to work in concert with plasmonic coupling in the process of $\pi-\pi$ interaction of a dye and nanoparticle assembly (Ravindran, Chandran, and Khan 2013).

Furthermore, the NPs of different size influence not only the surface plasmon wavelength but also the intensity of the electromagnetic field created in between the NPs leading to higher SERS enhancements (Ravindran, Chandran, and Khan 2013, Lim et al. 2006a).

XPS (X-ray photoelectron spectroscopy) is also the quantitative spectroscopic surface chemical analysis technique, giving access to qualitative, quantitative/semi-quantitative, and speciation information concerning the NPs surface (Lim et al. 2006a, Lazarević-Pašti et al. 2016). Under high vacuum conditions, X-ray irradiation of the nanomaterial leads to the emission of electrons, and the measurement of the kinetic energy and the number of electrons escaping from the surface of the nanomaterials gives XPS spectra. The binding energy can be calculated from kinetic energy and the specific groups of starburst macromolecules on the NPs surface can be identified and characterized.

DLS (dynamic light scattering) measurements represent the non-invasive, well-established technique for measuring the particles size distribution in order to characterize particles, emulsions or molecules, which have been dispersed or dissolved in a liquid (Bhattacharjee 2016). The Brownian motion of particles or molecules in suspension causes the laser light to be scattered at different intensities. Analysis of these intensity fluctuations yields the velocity of the Brownian motion and hence the particle size using the Stokes-Einstein relationship. The particle size, zeta potential, electrophoretic mobility, and conductivity are the key parameters for quantitative evaluation of particle stability in suspensions (Leroy, Tournassat, and Bizi 2011). The average NPs diameter (d_{av}) determined by DLS measurements represents the hydrodynamic diameter of the particle with hydration shell. Typical DLS spectra which show the distribution of intensity of scattered light as a function of the particle size distribution are shown in Figure 6.

However, DLS measures the hydrodynamic radius, i.e., the radius of the hypothetical sphere of the dispersed particles. It appears that the difference between these various methods represents the thickness of the diffuse layer (capping agent and solvation ions/molecules layer) on the particle corona.

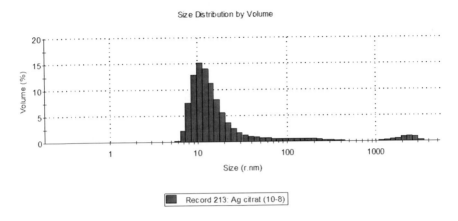

Figure 6. DLS diagrams for light scattering as a function volume on the diameter of citrate-capped AgNPs.

Besides, a polydispersity index (PDI) as a measure of heterogeneity of NPs size, suggests the degree of the homogeneity, since the zeta potential measurements indicate the charge of NPs surface, which strongly depends on noble metal reduction procedure and capping agents (Laban et al. 2016, Laban et al. 2014, Vujačić, Vasić, et al. 2013). Moreover, zeta potential values characterize the stability of colloid dispersions. It is the potential at the supposed slipping plane that separates the stationary and mobile phases in the tangential flow of the liquid with respect to the surface. Its value is closely related to suspension stability and particle surface morphology. The electric potential at the slipping plane is of particular interest in order to estimate the critical coagulation concentration when studying NPs agglomeration.

The surface conductivity is an important parameter for characterization of the electromigration of counter and co-ions of the electrical double layer along the surface of the particle. It is in accordance with zeta potential values and with literature, data indicate that this value is inversely proportional to the size of the NPs (Bhattacharjee 2016, Leroy et al. 2013, Leroy, Tournassat, and Bizi 2011).

The values of electrophoretic mobility parameters (Laban et al. 2014, Vujačić, Vasić, et al. 2013, Laban et al. 2016) are in correlation with the migration of charged colloidal particles depends on their size, zeta potential, surface conductivity and stationary medium (ionic strength and pH).

3.2.2. Nanomicroscopy Techniques

TEM (transmission electron microscopy) belongs the most important nanomicroscopy techniques for directly imaging nanomaterials (Alaqad and Saleh 2016). This technique shows the size of NPs within a different size range and illustrates their morphology image, shape, surface area, and the diameter Figure 7 represents AuNPs and AgNPs of various shape, size and surface capping.

Figure 7. TEM images of borate-capped spherical AuNPs (a), triangular Ag nanoplates (b) and CTAB capped Ag nanorods (c).

The analysis of TEM micrographs can also indicate the aggregation tendency as illustrated for assembly containing AuNPs and TC dye (Figure 8), which is also visible also in the slight change of the plasmon position in the absorption spectra of dye – NPs assembly. The reason for this behaviour is attributed to the molecules adsorbed on the NPs surface, which are involved in the interlinking of NPs. However, the question is how the partial charges of NPs and capping molecule overcome the electrostatic barriers in order to be tightly bound to each other.

AFM (atomic force microscopy) method is in general used to investigate the dispersion and aggregation of nanomaterials, in addition to their size, shape, sorption and structure (Alaqad and Saleh 2016, Zhang et al. 2016). Three different scanning modes are available, including contact mode, non-contact mode, and intermittent sample contact mode. As the example, Figure 9 represents AFM images of AgNPs in the absence and presence of TC dye. The topography of NPs indicates that they are nearly spherical in shape. Here, the advantage of this analysis is to produce images in all three dimensions, and the method for determining the diameter of NPs is to attach them to a flat substrate and measure the heights above this substrate. The AgNPs height (z-direction) at a distance where the tip is repelled or attracted by the forces due to the interaction with the surface was used to determine the diameter of bare NPs (Figure 9, c). The histograms obtained from the AFM images indicate the average diameters of particles. These results are higher than those obtained by TEM, but agree very well with DLS measurements (Vujačić, Vasić, et al. 2013, Vujačić, Vodnik, et al. 2013), as also obtained in the analysis of some similar colloid solutions. The high polydispersity index for colloid suspension indicates that the sample has a broad size distribution, which is also in good agreement with AFM measurements.

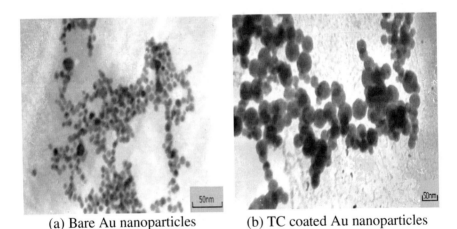

(a) Bare Au nanoparticles   (b) TC coated Au nanoparticles

Figure 8. The TEM analysis of (a) bare Au nanoparticles and (b) TC-coated Au nanoparticles.

Although the dye systems are fundamentally different from the conventional surfactant systems, the concentration-dependent morphological transition of supramolecular J aggregates of dyes can be observed using AFM (Laban et al. 2014, Laban et al. 2016). AFM topographical image and a cross-sectional profile of the string-like or sheet-like TC dye J aggregates (100–300 nm) are a complicated bundle of thin strings of 10–30 nm since in the sheet morphology the J-aggregates are stacked onto each other (Yao, Isohashi, and Kimura 2004, Bhattacharje 2016, Xu 2008).

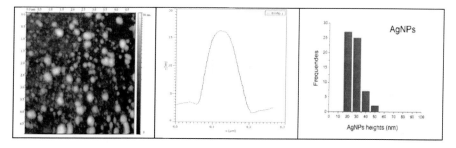

Figure 9. AFM images of topography (a), profile (b) and particle size distribution (c) of AgNPs.

3.3. Modification of AgNPs and AuNPs Surface with Thiacyanine Dyes

Modification of NPs surface by cyanine dyes leads to changes in their electronic properties (Burda et al. 2005, Chandrasekharan et al. 2000, Clark, Krueger, and Vanden Bout 2014, Hranisavljevic et al. 2002). The modulation of the chromophore's optical characteristics is obtained due to the electronic coupling of the dye exciton to the polarization of the metal NPs. However, some thiacyanine dyes form J-aggregates on the surface of silver and gold (NPs) which strongly depend on particle size and surface capping, as well on the structure of the dye (Kometani et al. 2001, Lim et al. 2006b, Neves and Andrade 2015). Self-organization of these molecules mediated by NPs is especially interesting because of the application of dye – NPs assemblies for nanoelectronics, medical diagnostics, drug delivery, chemical sensing and catalysis. The hybrid J-aggregate – NPs assemblies are also usually characterized by UV-vis spectrophotometry, TEM, AFM, DLS, FTIR, Raman spectroscopy, fluorescence measurements and DFT calculations (Ralević et al. 2018, Laban et al. 2016, Vujačić et al. 2012).

In general, some papers report the J-aggregation of thiacyanine dyes of similar structures in the presence of macromolecules, describing also its mechanism and kinetics (Avdeeva and Shapiro 2003, Chibisov, Slavnova, and Görner 2008, Voznyak and Chibisov 2008, Slavnova, Chibisov, and Görner 2005).

Table 2. Some specific characteristics of the AuNPs and AgNPs

Metal	PSD (nm)	Shape	Zeta potential (eV)	Surface covering	λ_{max} (nm)	Ref.
Au	6.0±0.5	sphere	-32.01±0.05	borate	520	(Vujačić et al. 2012)
Ag	6.0±0.9	sphere	-43.20±0.50	borate	388	(Laban et al. 2014)
Ag	10.1±0.9	sphere	-23.30±0.24	citrate	393	(Laban et al. 2016)
Ag	8x40	rods	24.22±0.95	CTAB	426 617	

3.4. Nanospectroscopy and Nanomicroscopy Methods for Dye-NPs Assembly Characterization

3.4.1. UV Vis Spectrophotometry and Fluorescence Spectroscopy

The basic methods to follow the formation of the thiacyanine dyes J-aggregates on the surface of NPs are UV Vis spectrophotometry and fluorescence spectroscopy.

3.4.1.1. Spectral Characteristics of TC Dye J-Aggregates

The absorption spectrum of TC dye (Table 1) shows a short-wavelength maximum at 409 nm denoted as a D-band which is assigned to the dimer (TC_2^{2-}) and a long-wavelength maximum (M-band) at 429 nm assigned to the monomer (TC^-) (Figure 10). The dye is most likely in aqueous solution present as an equilibrated mixture of monomers and dimers, similarly to other thiacyanine dyes studied by Chibisov and co-workers (Chibisov, Slavnova, and Görner 2008, Chibisov, Görner, and Slavnova 2004).

The monovalent and bivalent cations strongly promote J-aggregation of thiacyanine dyes in water solution, and also in colloid dispersion. In the case of TC, the sharp band with a peak at 464 nm in the presence of 1×10^{-3} M KCl is characteristic for its J-aggregates. Moreover, the cations also strongly influence the intensity of dye fluorescence spectra (Laban, Vodnik, and Vasić 2015, Laban et al. 2013). The other characteristic feature of J – aggregation is the strong enhancement of fluorescence intensity. This is illustrated in Figure 10, which shows the dramatic change of fluorescence spectra (green line) in the presence of various KCl concentrations. It is worthily to notice that the peaks at 464 nm and 468 nm are characteristic for J- aggregation effects on absorption and fluorescence spectra respectively, for the pure dye.

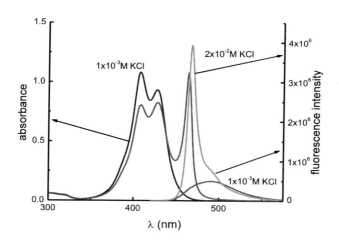

Figure 10. Absorption and fluorescence spectra of 1×10^{-5} M TC dye in the presence of KCl.

However, the emission band of J-aggregates is sharp and shifted to the red with respect to the absorption band by a Stokes shift, which is usually of the order of the line width but can also be vanishingly small in some systems. The fluorescence of J-aggregates can be excited either resonantly via the strong and narrow absorption J-band or non-resonantly via the relatively weak and broad absorption bands to the blue of the J-band. The excitations in such systems can relax either by incoherent energy transfer from some excited species (residual monomers, dimers, oligomers, etc.) to the emitting excitonic states of a particular segment of the J-aggregate or by intra-segment exciton energy conversion. As can be seen in Figure 10~~17~~, the narrow fluorescence band of TC J-aggregates (λ_{max} ~468 nm) is blue shifted and with dramatically increased intensity compared to the broad fluorescence band of dimmer (λ_{max} ~492 nm).

3.4.1.2. Influence of AgNPs and AuNPs on Absorption Spectra of TC Dye J-Aggregates

Some detailed studies deal with the photophysics of the TC J-aggregation on AgNPs and AuNPs surfaces in colloidal dispersions (Barooah et al. 2011, Chandrasekharan et al. 2000, Hranisavljevic et al. 2002, Jeunieau, Alin, and Nagy 1999, Kamalov, Struganova, and Yoshihara 1996, Lim et al. 2006b). The initial Au colloid dispersions exhibit an intense SPR peak at ~520 nm since AgNPs have the intensive plasmon peak at ~387 nm. The appearance of the J-aggregate exciton band was highly dependent on the NPs composition.

In the case of dye-coated Ag nanoparticles, the J-aggregate exciton band was observed in the form of a peak in the UV/vis spectrum (Liu, Sanyasi Rao, and Nunzi 2011, Laban et al. 2013). However, the presence of the TC J-aggregates on Au nanoparticles produced a sharp absorption minimum (dip) near the exciton resonance (Vujačić et al. 2012, Wiederrecht, Wurtz, and Hranisavljevic 2004). This was also found by TC, ThiaEt and PIC interaction with Au-nanorods having the spacer between the core and outer particle shell, as illustrated in some previous publications (Yoshida, Uchida, and Kometani 2009, Yoshida, Yonezawa, and Kometani 2009). These various spectral properties are due to differences in the coupling of dye J-aggregate excitons and polarizations in AuNPs and AgNPs. Specifically, polarization coupling to Ag at 475 nm involves only the intraband contribution from the surface plasmon of the metal core, while polarization coupling to Au involves contributions from polarizations derived from interband and intraband transitions. In the case of Au, the presence of the additional interband term leads to destructive interference between the exciton and plasmon polarizations.

From the spectrophotometric point of view, the formation of TC J-aggregates on the surface of NPs is followed by the appearing of a characteristic dip in the case of AuNPs or new sharp peak in the case of AgNPs, both at 481 nm (Figure 11) due to a coupling of J-aggregate excitons and polarizations in NPs. The positions of the dip and the peak are very close to the position of J – band of the pure dye (464 nm). However, in the peak type absorption spectra (AgNPs) J-band and plasmon appear independently in the absorption spectra of the composite, since in the dip type absorption spectra (AuNPs) there is the overlap between the J-band and the surface plasmon band. Moreover, dip – type absorption is the indication of the strong plasmon – exciton coupling, since the peak type absorption suggests usually the weak coupling. It is therefore reasonable that that the spectral line shape near the J-band can be used as a kind of measure for estimating the strength of exciton – plasmon coupling (Kometani et al. 2001).

Moreover, the stability and intensity of the absorbance in J-aggregation are strongly dependent on TC and NPs concentrations, time as well as on NPs type and surface covering.

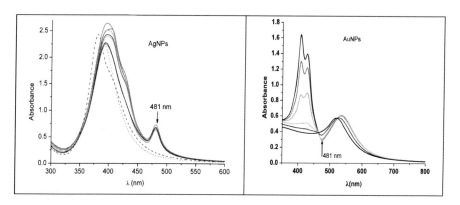

Figure 11. Absorption spectra of colloidal dispersions of 6 nm citrate-capped AgNPs (a) and AuNPs (b) in the presence of TC (0 - 1.67×10^{-5} M).

3.4.1.3. Fluorescence Spectra of NPs – TC Dye Assemblies

Thiacyanine dyes have the intensive fluorescence spectra and the fluorescence intensity is usually quenched in the presence of NPs regardless the J-aggregation occurs or not. As the example, the quenching fluorescence of TC dye depending on the concentration of 17 nm size citrate-capped AuNPs is presented in Figure 12. In this case, NPs do not induce the J-aggregation on their surface (Vujačić, Vasić, et al. 2013, Vujačić, Vodnik, et al. 2013).

The combination of fluorescence microscopy and optical spectroscopy gave birth to single molecule spectroscopy (SMS) as a method to study fluorescence light coming from a single molecule (or other nano-object) (Moerner and Kador 1989, Orrit and Bernard 1990, Scheblykin et al. 1996, Merdasa et al. 2014). SMS allows studying underlying dynamics that are not observable in experiments on bulk materials due to *ensemble averaging*. A well-known example is a fluorescence "blinking" as strong fluctuations of fluorescence intensity of a single molecule or single NP under continuous excitation, which is usually explained by the NP from time to time temporarily residing in a long-lived non-fluorescent "dark" state.

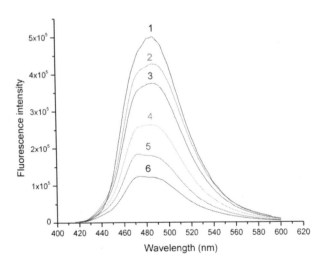

Figure 12. Fluorescence spectra of 1×10^{-6} M TC vs. AuNPs (17 nm) concentration from 1.9×10^{-10} M - 1.68×10^{-9} M.

The basic scheme of the setup is presented in Figure 13. Observation of fluorescence intensity blinking in multichromophoric systems (called also "collective blinking effect") was related to energy exchange between chromophores (Link and El-Sayed 1999, Liu, Sanyasi Rao, and Nunzi 2011, Lin et al. 2010).

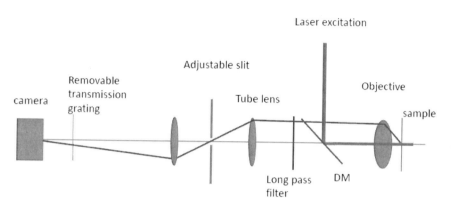

Figure 13. The basic scheme of single molecule spectroscopy (SMS) set-up.

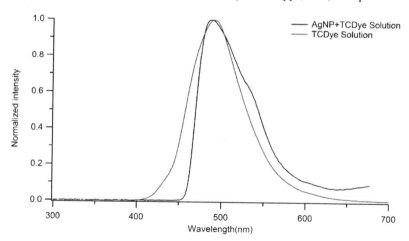

Figure 14. The emission of bulk solutions of TC dye in the absence and presence of AgNPs.

In the excited states of J-aggregates, the excitons are delocalized over a certain number of dye molecules depending on disorder and temperature (Struganova et al. 2003, West and Pearce 1965, Scheblykin et al. 2001).

Such a system consisting of coherently coupled chromophores could work as an extremely efficient light-harvesting antenna(Camacho et al. 2015). The SMS approach has been shown to be useful to study the structure of the exciton bands at low temperatures by monitoring excitation spectra of individual aggregates.

SMS has proven to be the most powerful tool to reveal underlying physical mechanisms behind the commonly observed ensemble-averaged properties of molecules, aggregates, and other nanosystems. Single molecular aggregate spectroscopy allows for extracting information about individual exciton levels (Brixner et al. 2017). It has been recently demonstrated that individual tubular J-aggregates immobilized on a surface can be studied experimentally. As the example, the emission of bulk solutions of TC dye in the absence and presence of AgNPs was measured with the excitation at different wavelengths and the results are presented in Figure 14. It seems however that AgNPs induced the J-aggregation, exerting the maximum at 492 nm and the shoulder at about 540 nm in the emission spectrum. For better clarification, the excitation at 458 nm was performed for increased TC concentration (Figure 15), and as the result the shoulder at 520 nm dramatically increased, confirming the observing an aggregate on AgNPs.

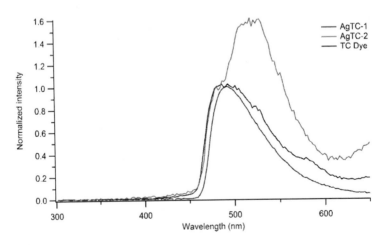

Figure 15. The emission spectra of TC - AgNPs assembly obtained due to the excitation at 458 nm for different TC concentrations.

The AgNPs surrounded by TC dye molecules are also observable at the level of the single molecules. Single molecule imaging of TC dye before and after background subtraction is presented in Figure 16. Surprisingly, the molecules are blinking and not very bright because of the inadequate camera settings available in this experimental setup, which gave the high background mainly from glass and other from impurities. This blinking is likely due to the quenching of the dye, but nevertheless, the emission and even blinking can be observed.

As a conclusion, the different wavelength available on the setup combined with emission spectroscopy seems to allow the distinguishing between the dyes in the monomer form and the J-aggregates. Different concentration of dyes and J-aggregates should minimize the amount of free dyes resulting in minimization of the background.

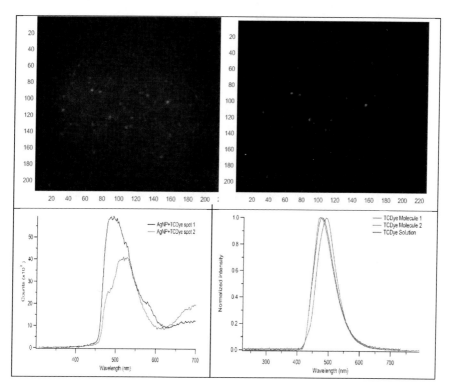

Figure 16. Single molecule imaging (upper images) and emission spectra (below) of AgNPs – TC assembly before and after the background subtraction.

3.4.2. TEM and AFM Analysis

TEM micrographs with particle size distribution (PSD) histograms for bare AuNPs and in the presence of TC are shown in Figure 25.

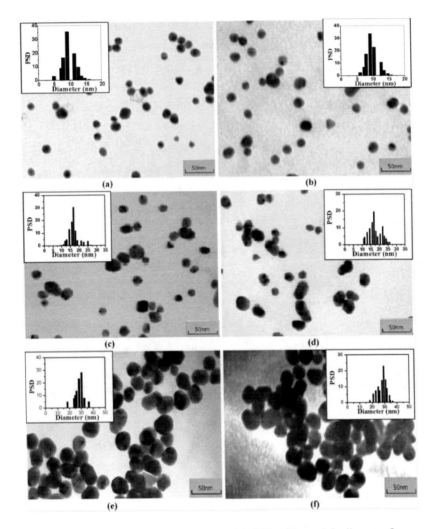

Figure 17. TEM images of bare and TC coated AuNPs with particle diameter 9 nm (a,b), 17 nm (c,d) and 30 nm (e,f), respectively.

The mean particle size of NPs was determined by fitting the obtained TEM data with Gaussian distribution function (Figure 17, inset). A TEM study of colloids in the absence and presence of TC confirmed the presence of nearly spherical shaped NPs, without the significant change of the core particle size diameter. However, the layer of film consisting of TC dye molecules capping their surface can be noticed (Vujačić, Vodnik, et al. 2013, Vujačić, Vasić, et al. 2013, Vujačić et al. 2012). It is worthily to notice that the dye is sorbed on the surface of AuNPs of selected particle size diameter without J-aggregation.

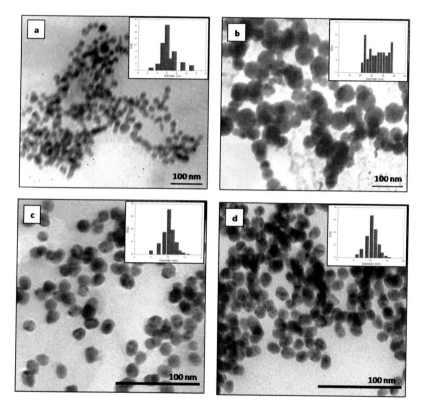

Figure 18. The TEM analysis of (a, c) bare Au nanoparticles and (b, d) TC-coated AuNPs of various sizes. Inset: particle size distribution (PSD).

Quite different interaction between AgNPs and TC dye was observed by analyzing TEM micrograms presented in Figure 18. Since the orientation of the particles is typically random, the resulting statistics are representative for the average PSD and standard deviation for the sample. The agglomeration of NPs was identified, which further indicated that TC dye sorption on the NPs surface induced their aggregation and cross-linking probably (Laban et al. 2014, Laban et al. 2016, Laban et al. 2013). This aggregation tendency was visible also in the UV-Vis spectrum of the dye – AgNPs assembly as a slight red shift of the LSPR position. The reason for this behaviour is attributed to TC molecules adsorbed on NP surfaces in J-aggregate forms, which are involved in the interlinking of metal NPs (Zhong et al. 2004).

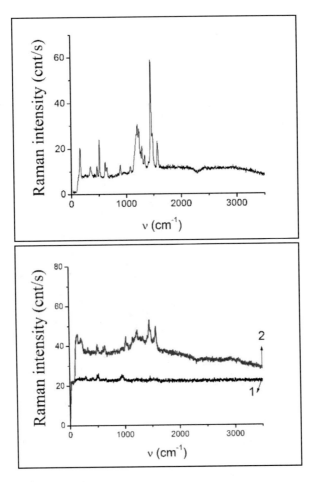

Figure 19. Upper panel: Raman spectra of solid TC dye. Bottom panel: Raman and SERS spectra of 5×10^{-6} M TC in the absence (1) and presence (2) AgNPs.

3.4.3. SERS and Raman Spectroscopy

Raman spectroscopy is a widely used method to analyze scattering of photons by molecules or molecular aggregates. The scattered light is then analyzed for frequency and polarization (Cheng et al. 2015, Kneipp et al. 2005, Kneipp et al. 2009). The formation of J-aggregates could occur on the surface of NPs, and the plasmonic coupling among NPs can be responsible for electrical field enhancement, which determines the optical and spectroscopic properties.

The SERS effect of TC in the presence of AuNPs is illustrated in Figure 19. The Raman spectra of solid dye show intensive bands in the region from 200 cm^{-1} - 1600 cm^{-1}, but there is no signal when present in solution at a concentration below $5x10^{-6}$ M. The addition of AgNPs into TC aqua solution induced SERS effect and produced the signal of significant intensity.

To obtain this spectrum a high laser power was needed. However, 5×10^{-6} M TC dye solution did not exhibit Raman peaks even at a laser power of ~200 $\mu W/\mu m^2$, but the addition of AgNPs into the solution induced SERS (Ralević et al. 2018).

3.5. Mechanism of J-Aggregation on NPs Surface

In the elucidation of the mechanism of dyes binding on the surface of NPs, they are usually considered as macromolecules with several binding sites and dye is considered as a ligand (Voet and Voet 1993, Reinheimer 1965, Munson 1984). Accordingly, the methods developed for the description of the interaction between biological macromolecules and ligands, such as Hill and Scatchard analysis (Scatchard 1949, Munson 1984, Gesztelyi et al. 2012) of saturation curves obtained using spectrophotometric and fluorescence data can be applied in order to evaluate the binding mechanism. These methods usually provide the data concerning the number of available binding sites per NPs, stability constants of ligand – NPs complex and also the information if the ligand binding is cooperative or random.

3.5.1. Fluorescence Quenching

The study of the dependence of TC dye fluorescence on NPs concentration offers the opportunity to clarify the mechanism of TC binding onto the NPs surface (Vujačić, Vasić, et al. 2013, Vujačić, Vodnik, et al. 2013, Laban et al. 2013, Laban et al. 2016, Laban et al. 2014, Vujačić et al. 2012). It was observed previously that the surface plasmon acted as an efficient energy acceptor for the NPs larger than 5 nm, even at the distance of 1 nm between the dye molecule and NPs surface (Lakowicz 2007). As a consequence, the dye - NPs assemblies usually exert concentration-dependent fluorescence quenching properties (Dulkeith et al. 2002, Dulkeith et al. 2005). However, it is worthily to notice, that it occurs regardless of NPs induce J – aggregation. Moreover, fluorescence quenching can evidently be ascribed to the energy transfer between TC dye and NPs due to the sorption of the dye on its surface.

3.5.1.1. Saturation Curves

As an example, the dependence of the fluorescence spectra of TC on the 10 nm particle size diameter citrate capped AuNPs concentration is presented in Figure 20 (Laban et al. 2016). The strong fluorescence band of 1×10^{-6} M TC colloid solution with the maximum at 490 nm is ascribed to TC fluorescence since AgNPs do not exert any significant fluorescence in this spectral range and under these experimental conditions (Siwach and Sen 2009). AgNPs induced the concentration-dependent decrease of the overall fluorescence intensity, as also observed in previous studies of TC interaction with Au and AgNPs (Vujačić, Vasić, et al. 2013, Vujačić, Vodnik, et al. 2013). In general, the fluorescence quenching can be ascribed to the formation of a nonfluorescent complex between the fluorophore and a quencher. The interaction between oxygen atom from SO_3^- group of TC dye and AgNPs surface resulted probably in complex formation (Ralević et al. 2018, Laban et al. 2016). The close distance between the fluorophore and the metallic NPs enabled the efficient energy transfer, which resulted in fluorescence quenching.

The quenching of the fluorescence is obvious with increasing of NPs concentration. Here, the constant TC dye concentration was applied followed by the variation of AgNPs concentration.

Assuming that the measured fluorescence can be ascribed to the unbound TC molecules in solution, the free and bound concentrations in the suspension can be easily calculated and presented as the dependence on AgNPs concentration. The dependences of free (unbound) and bound concentration of TC dye on AgNPs concentration are presented in Figure 20,b).

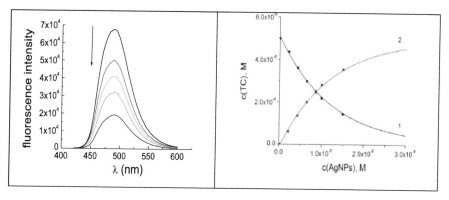

Figure 20. *a)* TC dye fluorescence spectra and *b)* dependence of free (1) and bound (2) TC on AgNPs concentration. TC – 5×10^{-6} M.

In order to describe the AgNPs - TC interaction, the Scatchard analysis is usually used for analyzing the data from freely reversible ligand/receptor interactions according to the relation:

$$(TC_{bound}/mol\ NPs)/TC_{eq} = NK_a - K_a(TC_{bound}/mol\ NPs) \qquad (1)$$

where TC_{bound} and TC_{eq} represent quenched and equilibrium TC concentration, N is the number of binding sites per NPs, and K_a is the apparent microscopic association constant. Here, AgNPs were considered as macromolecules with several binding sites and TC dye was considered as a ligand. The relation (1) is presented in Figure 21 and the results of Scatchard analysis give the values of the association constant, K_a and number (N) of TC molecules bonded per single AgNP. The Scatchard graph indicated that about 362 TC molecules were bonded per AgNP.

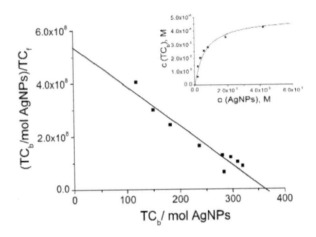

Figure 21. Scatchard analysis of saturation curve 2. Inset: Dependence of the concentration of bounded TC molecules on the concentration of free binding sites on AgNPs.

Further, the saturation curve was constructed as the dependence of concentration of bound TC molecules on the concentration of available free binding sites on AgNPs. For this purpose, the AgNPs concentration was multiplied by the number of binding sites obtained from Scatchard analysis and the number of bonded TC molecules was subtracted from the corresponding AgNPs concentration. Hill analysis of saturation curve (Figure 21, Inset) was performed according to the relation:

$$C_{TCb} = TC_{TCo} \times C^n/(K_a^{-n} + C^n) \qquad (2)$$

where C_{TCb} is the concentration bound ligand molecules, TC_{TCo} is yielding to the initial TC concentration in the colloid suspension, C is the concentration of available binding sites per AgNP, K_a is the affinity constant and n is the Hill coefficient (Gesztelyi et al. 2012). The value of n close to 1 suggests the noncooperative binding.

3.5.1.2. Stern – Volmer Graphs

The analysis of the experimental results using Stern–Volmer relation ((Lakowicz 2007) is presented in Figure 22. In these experiments, two TC concentrations were used for interaction with 6 nm particle diameter borate capped AgNPs. The Stern−Volmer relation accounting for both static and dynamic (collisional) fluorescence quenching is generally written as

$$I_0/I = 1 + K_{SV}(Q) \qquad (3)$$

where I_0 and I are the fluorescence intensities of TC in the absence and presence of AgNPs, Q is the quencher concentration, and $K_{SV} = K_S + K_D$ is the Stern−Volmer quenching constant (K_S and K_D are the static and dynamic quenching constants, respectively). The overlapping linear Stern−Volmer plot for the experimental data of two sets of measurements (5×10^{-6} and 1×10^{-6} M TC, the slope 2.35 ± 0.10, $R2 = 0.9935$) was obtained as presented in Figure 31. The linearity of Stern – Volmer plot indicates that only one type of quenching occurred. The high value of K_{SV} suggests that NPs quench the fluorescence with an extraordinarily high Stern–Volmer constant (K_{SV}) which is in the range of 10^8 M^{-1} in many cases.

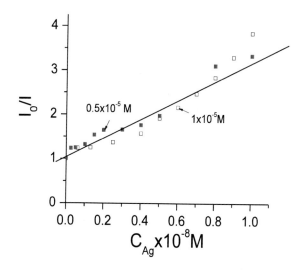

Figure 22. Stern Volmer graph for interaction of 6 nm size AgNPs with 1×10^{-5} M and 5×10^{-6} M AgNPs (Laban et al. 2016).

The very high value of the quenching constant indicates the presence of adsorption of the dye molecule on the AgNP surface and suggests a strong association between the nanoparticles and TC. The quenching of TC fluorescence occurred, but a change in the fluorescence band position in the presence of AgNPs was not observed. However, this cannot be the direct confirmation of the absence of significant molecular interactions under the prevailing experimental conditions, since it may be expected that the fluorescence of a bound TC molecule is subjected to complete quenching due to the TC interaction with AgNPs, such that only unbound TC molecules emit fluorescence.

3.5.2. Kinetics of J-Aggregation

J-aggregation kinetics, i.e., the dependence of J-aggregation formation on time, exhibits a various types of kinetic curves (Chibisov, Görner, and Slavnova 2004, Pasternack et al. 2000, Görner, Chibisov, and Slavnova 2006) depending on the dye structure, dye concentration, pH, presence of metal ions or NPs and temperature.

Kinetics of AuNPs and AgNPs mediated J-aggregation were studied in two modes: monitoring the change of absorbance or fluorescence. Figure 32 represents the data obtained for 6 nm size borate capped AuNPs which suggest that the absorbance at 475 nm (dip position) and the and fluorescence at 490 nm yielded very similar kinetic curves. The experimental data were best fitted to eq. 4 which is the approximated form of the equation describing two consecutive first-order reactions (Lewis 1974):

$$A = A_0 + A_1 \exp(-k_{obs1}t) + A_2 \exp(-k_{obs2}t) \tag{4}$$

where A_0, A_1, and A_2 are constants. The first step is ascribed to the fast direct adsorption of TC dye on the gold nanoparticles and includes the formation of J-aggregates of the adsorbed TC dye via $\pi-\pi$ interactions until sufficient surface coverage has been reached. This argument is strongly supported by the fact that both kinetic curves, one derived from the absorbance at 475 nm (Figure 23 a) and one derived from fluorescence at 490 nm (Figure 23,b) show this fast step. The 475 nm feature in the absorbance data represents direct evidence of J-aggregate formation on the surface of the Au nanoparticles. The second, slower process, again observed in both kinetic curves, can be ascribed to the growth of J-aggregates on the initial TC layer. In contrast to this simple two-step mechanism, the kinetics of thiacyanine J-aggregate formation in the absence of metallic nanoparticles occurs via a complex autocatalytic process, involving equilibrium between monomers and dimers of the dye (Vujačić et al. 2012, Vujačić, Vasić, et al. 2013, Chibisov, Görner, and Slavnova 2004).

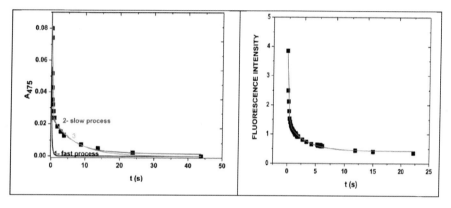

Figure 23. Change of absorbance (a) and fluorescence intensity (b) vs. time for the dispersion containing 1.67×10^{-5} M TC and 5×10^{-8} M AuNPs at 28 °C. (a) :1 - fast process; 2 – slow process; 3 – fit to kinetic eq.4.

The other types of kinetic curves, sigmoid in shape, are characteristic for reversible autocatalytic aggregation of dyes, pigments or other molecules(Chibisov, Görner, and Slavnova 2004, Pasternack et al. 2000, Kodaka 2004), while non-sigmoid type kinetic curves are analyzed in terms of a time-dependent rate constant and a simple exponential dependence. In general, the typical kinetic curves for the formation of J-aggregates with sigmoid shape could not be successfully fitted to standard first-order, second-order or coupled first-order equations. This type of kinetic curves is characteristic for TC J-aggregation on the surface of AgNPs of various size and surface capping.

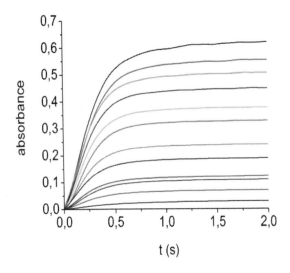

Figure 24. Kinetics of 1x10⁻⁵ M TC J-aggregation vs. AgNPs concentration from 0.25×10^{-8} M to 5×10^{-8} M.

A stretched exponential function (Görner, Chibisov, and Slavnova 2006, Pasternack et al. 2000, Johnston 2006) can be successfully applied to describe J-aggregate formation in systems containing independently relaxing species, each of which decays exponentially over time with a specific relaxation rate k_{app} (eq.5):

$$C_J = C_J^o + (C_J^{oo} - C_J^o)(1 - \exp(-(k_{app}t)^n)) \qquad (5)$$

Here, C_J^o and C_J^{oo} are the concentrations of J-aggregates immediately after mixing and after the completion of J-aggregation, respectively; k_{app} is the apparent rate constant of J-aggregates formation and n represents the stretch exponential parameter, i.e., the degree of the sigmoid character. In general, the stretched exponential function has usually been applied to a relaxation system in which the individual components are not independent and interact with each other. The obtained parameter k_{app} in the stretched exponential function represents the "average" relaxation rate for the particular experimental conditions. It is worth noting that the parameter n ranged from 0.90 to 1.11. This means that the sigmoid degree was not large, indicating a low degree of cooperation in the process of aggregate formation (Kodaka 2004).

The kinetic curve presented in Figure 24 is also characterized by an inflexion point. This has also been found in the studies of J-aggregation of some cyanine dyes in the presence of gelatin, as well as metal ions. The first derivative of the kinetic curve dC_J/dt *vs.* time has a bell-shaped form with the maximum value at the inflexion point. This value represents the maximum rate of J-aggregate formation.

In general, the reaction mechanism can be explained by a series of second- and first-order reactions, with a reversible first step according to the relation:

$$AgNPs + TC \underset{k_-}{\overset{k_+}{\leftrightarrows}} AgNPsTC \overset{k_2}{\rightarrow} J_{agg} \qquad (6)$$

It is reasonable to assume that the second-order reaction step corresponds to the adsorption of TC on the AgNP surface, and the first-order step corresponds to the formation of J-aggregates from the adsorbed TC molecules. As TC increases, the AgNPs surface is completely occupied and further addition of TC does not affect the rate. In other words, the rate-determining step has now shifted toward a first-order reaction of the formation of J aggregates. It can be easily shown that the following dependence of k_{app} on the TC concentration can be derived:

$$k_{app} = k_2(TC)/(K_m + (TC)) \qquad (7)$$

where $K_m = (k_2 + k^-)/k^+$ represents the pseudo-equilibrium Michaelis constant. Here, k^+ and k^- are the forward and backward rate constants of the first reaction step in eq 5. However, at low TC concentrations, the rate-determining step can be considered as the formation of an intermediate, i.e., the adsorbed TC molecules on AgNPs. The formation of J-aggregates followed a low first-order rate constant, and k_{app} tended toward the maximum value of k_2 as TC increased.

4. Theoretical Studies of Cyanine Dyes J-Aggregation

4.1. Optical Band Shapes Induced by H- and J-Aggregation

A deep understanding of the dye-aggregation mechanism is very important for the design of new dyes, detailed atomistic understanding of the functioning of the related devices and for the behaviour of the aggregates in a biological environment. The information about the aggregation of polymethine dyes can be often obtained from the shape of the optical absorption bands, leading to the conclusion about the entire structure of the aggregate and the number of molecules in the optical chromophore and in the aggregate (Egorov 2017, 2009, Egorov and Alfimov 2007).

There is a number of theoretical methods, such density functional theory(DFT), hybrid DFT functional M06-2X, complete active space second-order perturbation theory (CASPT2), quantum Monte Carlo methods (QMC), coupled cluster linear response up to third approximate order (CC3), various flavors of time-dependent density functional theory (TDDFT), including the recently proposed perturbative correction scheme (B2PLYP) and second-order Møller-Plesset perturbation theory (MP2) which have been carried out to investigate how different modes of aggregation lead to changes in the UV-VIS spectra of these molecules (Pastore and De Angelis 2010, Majhi et al. 2014, Egorov 2017).

The band shapes in the absorption spectra of dimers can be simulated using a combination of an empirical molecular force field (CNDO/S) for the ground state with quantum chemical calculations of the electron excitation energy as a function of mutual shift of chromophores relative to each other in order to obtain the electronic and optical properties of molecular aggregates of the dye, as well as their structures from molecular dynamics (MD) simulations of the self-aggregation process (Egorov 2017, 2009, Egorov and Alfimov 2007, Haverkort et al. 2013).

Semiempirical MD simulations for amphi-PIC using GROMOS53A6 force field, with additional parametrizations calculated by B3LYP 6-31G* for a central bond, were performed by Haverkort et al. Simulation box consisted of equal number (80-160) of positive amphi-PIC ions and negative counterions and a large number of water molecules. Depending on the initial concentration of amphi-PIC molecules in our MD simulations of spontaneous aggregation, aggregates with different geometries were formed within simulation time of a few hundreds of nanoseconds, including a single-walled cylinder or a ribbon. The internal structure of these aggregates using the π-π stacking and the average orientation of the long axis of the amphi-PIC molecule's chromophore indicated that the molecular arrangement exhibited much disorder which is in accordance to the wide absorption band observed for aggregates of amphi-PIC, since the changing the counterion of the positively charged amphi-PIC dye can change the equilibrium aggregate shape. The structural information from MD simulation was translated into an effective time-dependent Frenkel exciton Hamiltonian for the dominant optical transitions in the aggregate, to further calculate the apsorption spectrum. Excitonic couplings and molecular transition energies were calculated by quantum chemical methods. The absorption spectra of the aggregate structures obtained from MD feature a blue-shifted peak compared to that of the monomer, implying that H-aggregates were obtained. Comparison to the experimental absorption spectrum of amphi-PIC aggregates shows that the simulated line shape is too wide, pointing to too much disorder in the internal structure of the simulated aggregates, and implying to the additional factors determining aggregate behavior that were not included in the MD simulation (Haverkort et al. 2013, Haverkort et al. 2014).

The shape of absorption bands of aggregates formed by two, four, and nine molecules of a polymethine dye was calculated by the Monte-Carlo method. The energy of interaction of the molecules in the ground state was simulated using atom-atom potentials, and the energies of interaction between dipole moments of electronic transitions of the monomers were estimated by quantum-chemical methods. In the dimer aggregate, the dipole moments of the electronic transitions in the monomers interact weakly; therefore, the electron absorption spectrum should be similar to that of the monomer. Going from the dimer to the aggregates consisting of four and nine monomers, the relative positions of monomers change and this, in turn, increases the energy of interaction between the dipole moments of their electronic transitions, resulting in a red shift characteristic of J-aggregates and narrowing of the absorption bands (Burshtein, Bagaturyants, and Alfimov 1997).

Theoretical treatments of the optical band shapes for monomers (M) of polymethine dyes, their dimers (D) and J-aggregates, as well as M–D and M–J concentration equilibria, were previously given by V.V. Egorov (Egorov 2017) based on the so-called dozy-chaos theory of molecular quantum transitions. Although there are some nanomicroscopy methods with the opportunity to observe aggregates, spectroscopy still remains the main method to obtain information about them.The diversity of theoretically obtained optical band shapes of aggregates of thiapolymethinecyanine dye obtained by applying dozy chaos theory, including monomers, dimmers, J- and H-aggregates, were compared to the available spectroscopy experimental data.

4.2. Theoretical Prediction of Thiacyanine Dyes J-Aggregation on NPs Surface

The J-aggregates of cyanine dyes formed by the dyes deposited on plasmonic gold and silver NPs induce various optical effects (Patra, Chandaluri, and Radhakrishnan 2012, Fofang et al. 2008). NPs generate localized plasmons that provide strong modulation of the light absorption and scattering showing the features of coherent coupling between the localized plasmons of the metallic nanoparticle and the excitons of the molecular J-aggregate. Such coherent coupling was demonstrated in experiments with gold and silver tunable plasmonic nanoparticles covered with inert molecular layers of variable width.

More than 20-fold plasmonic enhancement of J-aggregates fluorescence could be expected if the experimental conditions become optimal. It should be noted that the enhancement effect also depends on the J-aggregate structure: the strongest effect is expected from J-aggregates with large exciton coherence length.

To control the distance between J-aggregates and AgNPs the latter was covered by a polymer shell of variable thickness using the layer-by-layer assembly method.

The theoretical calculations, compared to the experimental results, confirmed that the position of peaks in the extinction spectra of the NPs depends on their geometric parameters and the optical constants of the core and shell. The results showed the dependence of the calculated absorption cross-section on the thickness of the TC J-aggregate layer.

Some recent publications are devoted to the theoretical studies of the light absorption and scattering cross-sections of metallic (Ag, Au, Cu and Al) NPs coated with different cyanine dyes (Lebedev and Medvedev 2013a, Lebedev and Medvedev 2013b). The purposes were to construct a theoretical model capable of explaining and quantitatively describing the results obtained by measuring in detail absorption spectra of J-aggregate coated metal NPs. The two-component metal/J-aggregate NPs with various geometric parameters containing TC, OC (3,3'-disulfopropyl-5,5'-triethylammonium salt) or PIC have the absorption peaks in different spectral regions relative to their plasmon resonance peaks and differ drastically in the oscillator strength of transitions in their J-band (Lebedev and Medvedev 2013b, Lebedev and Medvedev 2013a, Lebedev et al. 2008). Moreover, they differ qualitatively in the nature of the interaction of a Frenkel exciton with localized plasmons and also in the influence of this interaction on the spectral characteristics of the nanosystems in the weak and strong plasmon-exciton coupling regimes. The calculated absorption peaks in the J-band of such dyes are situated in different spectral regions relative to the plasmon resonance peaks of Ag and Au particles, and the dyes differ drastically on the particle core radius were and the shell thickness consisting of J-aggregates of day.

The characteristic J-aggregate peaks differ in position or height because of different effects in plasmon-exciton coupling in such systems, as well as the outer radius of the particle. The polarizability of the hybrid system depends not only on the permittivity of the core and shell but also on the ratio of the inner and outer radii of the concentric spheres induced by the electromagnetic core-shell coupling (Lebedev et al. 2008). Therefore, varying this ratio allows peak positions and heights in the extinction spectrum of composite particles to be tuned over rather wide ranges. he absorption spectrum calculated for a metal-organic (Ag/J-aggregate) NPs suspension with allowance for the particle size distribution and the experimental absorption data for the NPs in suspension showed that the calculation results agree well with the experimental data. Based on the obtained results, an approach to controlling the optical properties of the metal-organic nanoparticles can be proposed.

4.3. Adsorption of TC-Dyes on the NPs Surface - An Ab Initio Approach

A density functional theory (DFT), time-dependent DFT, and *ab initio* second order Møller−Plesset perturbation theory were used to study of the aggregation of the metal free indoline D102 and D149 dyes on extended TiO_2 model clusters (Pastore and De Angelis 2010). The ability of dyes to form aggregates was demonstrated by MP2 calculations, showing that a dimer is more stable than two isolate monomers for about 9 kCal/mol, while DFT did not predict dye aggregation as energy decreased with increasing distance between monomers. It was found that the bidentate adsorption mode via oxygen atoms is largely energetically favored compared to the monodentate one, by as much as 30 kcal/mol. Such result implies that the dye molecules are lying almost flat with respect to the surface, rather than pointing in an orthogonal direction to the surface plane. Excited state calculations on the selected dimers almost quantitatively reproduced the measured red-shifts, predicting values of 0.15 and 0.08 eV for D102 and D149, respectively, confirming the overall picture extracted from the adsorption studies. This computational study was the first attempt to fully describe the behavior of TC dyes on some support using only ab initio methods.

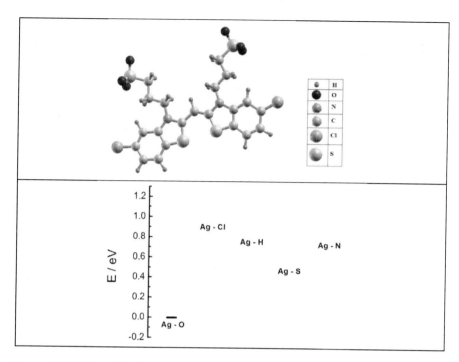

Figure 25. DFT-optimized geometry of TC dye anion visualized with XcrysDen Software (upper panel) and relative energies of Ag interactions with TC-dye functional groups (bottom panel).

The simple semiquantitative model to determine DFT interaction energy of different binding sites of 3,3′-Disulfopropyl-5,5′-Dichlorothiacyanine with one Au or Ag atom was proposed by Laban et al. (Laban et al. 2016). The set of DFT-LDA calculations was used, considering NPs surface as a single atom, interacting with different parts of the dye molecule anion (Laban et al. 2016, Ralević et al. 2018, Milović et al. 2019, Vasić Anićijević et al. 2015). The obtained results for DFT optimization of the TC dye anion geometry revealed a rather planar geometry (Figure 25), which is in good agreement with previously obtained results.

Figure 25 represents the energies of binding between Ag atoms from different functional groups of the TC dye and Ag atoms. The most stable configuration through sulphonate oxygen (Ag-O) is taken as a reference point. The obtained results suggest that partially negative oxygen from SO_3^-

groups exhibits the strongest interaction with the Ag atom. Similar results were obtained earlier in an *ab initio* study of oxygen adsorption on semiconducting surfaces (Vasić Anićijević et al. 2016, Vasić Anićijević et al. 2015, Lazarević-Pašti et al. 2018, Maslovara et al. 2019, Ralević et al. 2018). Such an adsorption mode also allows the dye to be adsorbed in a variety of tilted geometries, as demonstrated for the example of a small Ag-cluster (Figure 26), which is in good agreement with the results of spectrophotometric and fluorescence studies, and also indicated a slanted orientation of the dye on the NP surface.

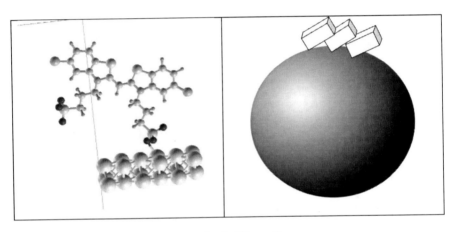

Figure 26. Possible TC orientation on the AgNPs surface.

Model was widened to the adsorption on 18-atom Ag-cluster in order to compare adsorption energies of TC dye and AgNP capping citrate and borate anions. It was shown that the adsorption of 3,3′-Disulfopropyl-5,5′-Dichlorothiacyanine is facilitated in presence of K^+ ions, since the deposition of K+ ions on the surface inevitably weakens the adsorption of capping (citrate or borate) anions, thus providing the termodynamical driving force for their replacement by TC dye anions. This study represents the first attempt to understand interaction between TC dye and metallic NP core in a complex real environment using ab initio methods, implying that adsorption of TC dyes on metallic NPs is significantly dependent on surrounding ions. The results are further supported by subsequent SERS findings (Ralević et al. 2018) that citrate-capped AgNPs are only partially covered by TC dye, while borate capped NPs are fully covered, being in a good agreement with DFT finding that citrate is more strongly bound to the NP surface than borate.

The similar one-metal-atom model in combination with DFT -CAM-B3LYP was used to study adsorption of piroxicam on Au surface (Binaymotlagh et al. 2017), showing that the interaction energy with Au-atom is the highest for N-atom of pyridil ring.

In addition, it was shown that including more than two surface layers do not have a significant effect on the geometry of the adsorbed dye on the surface. The reason for this can be attributed to the distance between the atomic layers of the Au and Ag surface. So in geometry optimization only topmost layer of Au atoms was treated by quantum mechanics and two bottom layers were treated by semiempirical methods During geometry optimization orientation of the molecule changed from horizontal to perpendicular, revealing that the (C–O)amide, S–O, and pyridyl moieties can simultaneously take part in the interaction of Pyr with the Au NPs (Binaymotlagh et al. 2017).

5. APPLICATION OF J-AGGREGATES

The self-organization of thiacyanine dyes in solution or at a solid-liquid interface is a very interesting behaviour of these systems. Their aggregation in solution has some analogous characteristics to aggregates important in biology, such as those in photosynthetic systems. In addition, aggregation of cyanine dye molecules in solution in the vicinity of metal NPs has captured recent interest due to the ability of coupling between dye aggregates and metal nanocrystals. The effects associated with close proxymity of the dye to the metal surfaces can result in shifting the SPR of the metal NPs and/or conversely, can result in enhanced emission in surface-enhanced Raman scattering (SERS) or enhanced/quenched fluorescence of the dye.

5.1. J-Aggregates as Photosensitizers in Photonic Devices

One among the first applications of cyanine dyes J-aggregates was in silver halide film photography due to their intense light absorption and strong reducing power for metal cations (Tani 2011). The advantages of using J-aggregates, over monomeric dyes are their enhanced wavelength-selective absorbance and also ultrafast energy and electron migration that dramatically increase the efficiency of latent image formation. Due to these intriguing properties, the J-aggregates are usually called 'supersensitizers'. For example, the mechanism of dye sensitization operates in the reduction of silver halides and electron transfer to a conduction band of a semiconductor, such as for instance TiO_2, in dye-sensitized solar cells, where the advantage of J-aggregates as supersensitizers can be efficiently explored (Chen et al. 2005).

In organic solar cells, the ability of J-aggregates to display high absorbance and exciton migration across long distances is the great advantage, because the light harvesting and sensitization need very efficient light absorbers that are also able to provide fast and efficient unidirectional electron transfer reaction to nanoporous surface of a metal oxide electrode (Saccone et al. 2016). The thin film formed of J-aggregates of only several nanometers possess a very large peak absorbance and can absorb and transduce the energy of light. Based on these features, it may become an ideal component of efficient low-cost artificial light-harvesting units (Wang and Zhan 2016).

The strongly wavelength-selective absorption of J-aggregates adsorbed on metal oxide thin film can also be applied in photodetectors. The construction of hybrid structures for colour imaging oroptical signalling extends their use in IR spectral range. Further, the inclusion of cyanine dyes in the form of J-aggregates into electron-hole conducting polymers results in efficient electroluminescence in single-layer light-emitting diodes that can provide a very narrow emission band with a maximum in the near-IR region. It must be pointed out that there is a strong difference between the interaction of cyanine dyes and conducting polymers leading to J-aggregation on solid polymer surface and in solutions (Osedach et al. 2012).

The application of fluorescent J-aggregates to chemical sensing and biosensing offers the new prospective solutions to these technologies that are in great demand. This application is based on the J-aggregate formation/disruption, the superquenching and the operation of smart FRET-based composites (Demchenko 2013).

The dramatic response in absorption and/or fluorescence spectra generated upon the dye aggregation/deaggregation or the transition between H-type and J-type aggregates is usually induced or assisted by the target. Here the target binding just shifts the equilibrium between monomeric and aggregated forms, and this may be enough for determining the targets based on extremely strong spectroscopic changes. Such switching of intensity between separate bands is especially valuable since this allows high precision in wavelength-ratiometric fluorescence detection (Bricks et al. 2017).

In biosensing, the targets that can be recognized by modulation of dye aggregate structures are various biomacromolecules, including DNA, polysaccharides and proteins (Demchenko 2010, Tatikolov 2012, Hannah and Armitage 2004). Owing to their very pronounced spectral and fluorescence properties, J- aggregates show the great abilities for the application in the chemical and biological sensing and represent the new prospective solutions in these technologies. The sensing is based on the dramatic response in absorption and fluorescence spectra due to the interaction dye-target molecule, that can be generated upon the dye aggregation/deaggregation or the transition between H-type and J-type aggregates induced or assisted by the target. Here the target binding just shifts the equilibrium between monomeric and aggregated forms, which induce strong spectroscopic changes. Cationic cyanine dyes can form J-aggregates on complexation with a variety of biomacromolecules, such as DNA, polysaccharides, and protein macromolecules. The specificity of target recognition by modulation of the dye aggregate structure can be increased by particular dye design.

The negative aspect of the J-aggregate application in routine fluorescent detection must also be taken into account, since the formation of the aggregates on certain biological structures is much more sensitive to the experimental conditions (i.e., biological molecule conformation, membrane phase, biological object and dye concentrations, ionic strength, etc.) than the dye monomer binding to the target species.

5.2. Application in Biological and Medical Imaging

The brightness, the wavelength-selectivity, the chemical stability and photostability are the major factors needed for successful application of J-aggregates in cellular microscopy for biological and medicinal imaging. However, the brightness and the wavelength-selectivity are easily achievable demands, but in order to achieve the chemical and photostability a lot of work is needed to satisfy them. The success in this challenging work is hard to predict, but it may be expected that it will be connected with the design of ultra-stable multifunctional nanoparticles and nanocomposites.

Despite their unique optical properties, there have been very few applications of J-aggregates for cellular and in vivo imaging (Jones et al. 2001, Bahmani et al. 2014, Jung, Vullev, and Anvari 2014). The high quantum yield of fluorescence emission is essential for applications in microscopy. The most attractive here is the near-IR absorbing J-aggregates that are able to be excited by light penetrating deeply into tissues and to form images on the background of low light scattering. A variety of cyanine dyes were designed for staining the membranes of living cells, containing acyl chains that allow their incorporation into the membranes with the charged fluorophore located among polar lipid heads and the tails extending into membrane interior. In some cases, these dyes can form J-aggregates, but their formation depends on the dye structure and also on the composition of lipids and their own lateral motion. Besides, the utility of J-aggregates has been hampered by low stability of their structures and the dependence of optical properties on their packing arrangement that may vary in biological environments.

It is well known that indocyanine green (ICGD exhibits concentration-dependent J-aggregation that depends on binding to proteins and lipids (Bahmani et al. 2014). Thus, it switches from an aggregated to a monomeric state upon binding to serum albumin leading to concomitant changes in absorption and fluorescence. Stability of its aggregates and their prolonged circulation in vivo (which is commonly very short, 2–4 min) can be increased by the formation of nanoparticles. In this way, by modification of nanoparticles surface, their targeted accumulation in cancer tissues can be achieved.

Two-photon fluorescence (TPF) microscopy is also a powerful tool for biological imaging and sensing applications, where the use of near-IR laser light allows forming sharp focal plane together with avoiding auto-fluorescence from the biological background (D'Avino, Terenziani, and Painelli 2006). Recently, the two-photon excited fluorescence in the aqueous dispersion of J-aggregates of anionic cyanine dye stabilized due to the interaction with cationic polymer poly (allylamine hydrochloride) was observed. The stabilization of formed J-aggregates induced sharp absorption band at 460 nm. Thus, the two-photon excitation in the near-IR range is prospective for tissue imaging in vivo with the benefits of near-transparency of many tissues in this spectral range and the reduction of undesired photo-damage of studied tissues.

The application of J-aggregates in super-resolution fluorescence microscopy techniques (nanoscopy) is based on their collective fluorescence blinking ('superblinking'), connected with the long-range exciton migration (Lin et al. 2010). Here, the high photostability of fluorescence emitters is strongly desirable. In contrast to the monomeric form of the cyanine dyes which are easily photodegradable, the J-aggregates seem to be much more photostable, which is supported by various results. For example, in the aerated polymer films, their photobleaching is low and depends on the type of polymer. The increased photostability was achieved for J-aggregate absorbed on nanoparticles (e.g., TiO_2) after the covering its layer with polysilanes.

CONCLUSION

The polymethine group of synthetic compounds (cyanine dyes), has applications in a variety of spectroscopy detection techniques in many fields of science and technology (to increase the sensitivity range of photographic emulsions, to label proteins, antibodies, peptides, nucleic acid probes and any kind of other biomolecules). They undergo J- and H-aggregation. This process is the spontaneous self-organization of the dye molecules in a parallel way (plane-to-plane stacking) to form a sandwich-type arrangement (H-aggregates) with a blue-shifted absorption band or a head-to-tail arrangement (end-to-end stacking) to form J-aggregates with a red-shifted absorption band in the absorption spectrum with respect to the monomer absorption.

Metal NPs, with their unique optical properties (size and surface capping dependent LSPR), support this process. Self-organization of thiacyanine dyes mediated by NPs is especially interesting because of the application of dye – NPs assemblies for nanoelectronics, medical diagnostics, drug delivery, chemical sensing and catalysis. Two typical aggregation behaviours of thiacyanine dyes, i.e., H- and J-aggregation, can be selectively formed via the proper selection of solvents, resulting in a significant difference in material properties and the resultant photovoltaic performance. Therefore in dye/NPs bi-layer hetero-junction photovoltaic cells, a significant contribution of J-aggregation to the photocurrent and a deeper ionization potential could be observed, most likely due to the intermolecular charge transfer interaction, which thus gave rise to larger open circuit voltages than those of H-aggregation based devices. The J-aggregation formation shall be considered as an effective approach to tune the optical/electrical properties in organic optoelectronic devices.

For the characterization of dye self-organization, a great number of nanospectroscopy techniques are widely used. These are UV-vis, Raman and fluorescence spectroscopy, TEM, AFM, Fourier transform infrared spectroscopy (FTIR), DLS, or DFT calculations, which also offer insights to the mechanism and rate of this process. The mechanism of interaction can be elucidated by application of the methods considering NPs as macromolecules and dyes as ligands. DFT calculations help to clarify the nature of dye binding to the NP surface. Using the theoretical model, the binding energy between

NPs and atoms from the dye structure can be calculated. These results enable to predict the orientation of the dye on the NPs surface.

The mechanism of TC – NPs interactions can be successfully understood by an examination of the changes in the absorption and fluorescence spectra. The quenching of the dyes' fluorescence is quantitatively related to the surface coverage of the dyes on the nanocrystal surfaces. Kinetic measurements provide important information for assessing a two-step process involving fast adsorption of the dye on the NPs surface in a combination with a slower process – the growth of J-aggregates on the initial TC layer. Finally, knowing the mechanism of self-organization processes (J- and H-aggregation) enables scientists to compose dye – NPs assemblies with the desired properties for applications in various fields of science and technology, especially in the photographic industry and as the sensors in medicine and environment.

ACKNOWLEDGMENTS

The Authors would like to thank the Ministry of Education and Science of the Republic of Serbia (Project No. 172023, 172045 and172056) for their financial support. Financial support by the European Cooperation in Science and Technology through COST Action MP1302 Nanospectroscopy is gratefully acknowledged. TLP is grateful to Prof. Ivan Scheblykin, Chemical Physics Division, Lund University in Lund, Sweden, for experimental work performed in his laboratory during her STSM.

REFERENCES

Alaqad, K., and T. A. Saleh. 2016. "Gold and Silver Nanoparticles: Synthesis Methods, Characterization Routes and Applications towards Drugs." *Journal of Environmental and Analytical Toxicology* no. 6 (4):384. doi: 10.4172/2161-0525.1000384.

Avakyan, V. G., B. I. Shapiro, and M. V. Alfimov. 2014. "Dimers, tetramers, and octamers of mono- and trimethyne thiacarbocyanine dyes. Structure, formation energy, and absorption band shifts." *Dyes and Pigments* no. 109:21-33. doi: 10.1016/j.dyepig.2014.04.026.

Avdeeva, V. I., and B. I. Shapiro. 2003. "J-Aggregation of Cyanine Dyes in Gelatin Solutions and Matrices." *Doklady Physical Chemistry* no. 389 (1-3):77-79. doi: 10.1023/A:1022962509897.

Bahmani, B., Y. Guerrero, D. Bacon, V. Kundra, V. I. Vullev, and B. Anvari. 2014. "Functionalized polymeric nanoparticles loaded with indocyanine green as theranostic materials for targeted molecular near infrared fluorescence imaging and photothermal destruction of ovarian cancer cells." *Lasers in Surgery and Medicine* no. 46 (7):582-92. doi: 10.1002/lsm.22269.

Barooah, N., A. C. Bhasikuttan, V. Sudarsan, S. D. Choudhury, H. Pal, and J. Mohanty. 2011. "Surface functionalized silver nanoparticle conjugates: demonstration of uptake and release of a phototherapeutic porphyrin dye." *Chemical Communications* no. 47 (32):9182-9184. doi: 10.1039/C1CC12354H.

Behera, G. B., P. K. Behera, and B. K. Mishra. 2007. "Cyanine Dyes: Self Aggregation and Behaviour in Surfactants A Review." *Journal of Surface Science and Technology* no. 23 (1/2):1.

Benson, R. C., and H. A. Kues. 1977. "Absorption and fluorescence properties of cyanine dyes." *Journal of Chemical & Engineering Data* no. 22 (4):379-383. doi: 10.1021/je60075a020.

Bhattacharjee, S. 2016. "DLS and zeta potential - What they are and what they are not?" *Journal of Controlled Release* no. 235:337-351. doi: 10.1016/j.jconrel.2016.06.017.

Binaymotlagh, R., H. Farrokhpour, H. Hadadzadeh, S. Z. Mirahmadi-Zare, and Z. Amirghofran. 2017. "Combined Experimental and Computational Study of the In Situ Adsorption of Piroxicam Anions on the Laser-Generated Gold Nanoparticles." *The Journal of Physical Chemistry C* no. 121 (15):8589-8600. doi: 10.1021/acs.jpcc.6b12962.

Bricks, J. L., Y. L. Slominskii, I. D. Panas, and A. P. Demchenko. 2017. "Fluorescent J-aggregates of cyanine dyes: basic research and applications review." *Methods and Applications in Fluorescence* no. 6 (1):012001. doi: 10.1088/2050-6120/aa8d0d.

Brixner, T., R. Hildner, J. Köhler, C. Lambert, and F. Würthner. 2017. "Exciton Transport in Molecular Aggregates – From Natural Antennas to Synthetic Chromophore Systems." *Advanced Energy Materials* no. 7 (16):1700236. doi: 10.1002/aenm.201700236.

Burda, C., X. Chen, R. Narayanan, and M. A. El-Sayed. 2005. "Chemistry and Properties of Nanocrystals of Different Shapes." *Chemical Reviews* no. 105 (4):1025-1102. doi: 10.1021/cr030063a.

Burshtein, K. Y., A. Bagaturyants, and V. M. Alfimov. 1997. *Computer simulation of the shape of absorption bands in electronic spectra of J-aggregates*. Vol. 46.

Camacho, R., S. Tubasum, J. Southall, R. J. Cogdell, G. Sforazzini, H. L. Anderson, T. Pullerits, and I. G. Scheblykin. 2015. "Fluorescence polarization measures energy funneling in single light-harvesting antennas—LH2 vs conjugated polymers." *Scientific Reports* no. 5:15080. doi: 10.1038/srep15080.

Chandrasekharan, N., P. V. Kamat, J. Hu, and G. Jones. 2000. "Dye-Capped Gold Nanoclusters: Photoinduced Morphological Changes in Gold/Rhodamine 6G Nanoassemblies." *The Journal of Physical Chemistry B* no. 104 (47):11103-11109. doi: 10.1021/jp002171w.

Chen, X., J. Guo, X. Peng, M. Guo, Y. Xu, L. Shi, C. Liang, L. Wang, Y. Gao, S. Sun, and S. Cai. 2005. "Novel cyanine dyes with different methine chains as sensitizers for nanocrystalline solar cell." *Journal of Photochemistry and Photobiology A: Chemistry* no. 171 (3):231-236. doi: https://doi.org/10.1016/j.jphotochem.2004.10.016.

Cheng, H. W., S. I. Lim, W. Fang, H. Yan, Z. Skeete, Q. M. Ngo, J. Luo, and C. J. Zhong. 2015. "Assessing Interparticle J-Aggregation of Two Different Cyanine Dyes with Gold Nanoparticles and Their Spectroscopic Characteristics." *The Journal of Physical Chemistry C* no. 119 (49):27786-27796. doi: 10.1021/acs.jpcc.5b09973.

178   *D. Vasić-Aničijević, T. Lazarević-Pašti and V. Vasić*

Chibisov, A. K., H. Görner, and T. D. Slavnova. 2004. "Kinetics of salt-induced J-aggregation of an anionic thiacarbocyanine dye in aqueous solution." *Chemical Physics Letters* no. 390 (1–3):240-245. doi: 10.1016/j.cplett.2004.03.131.

Chibisov, A. K., T. D. Slavnova, and H. Görner. 2008. "Self-assembly of polymethine dye molecules in solutions: Kinetic aspects of aggregation." *Nanotechnologies in Russia* no. 3 (1-2):19-34. doi: 10.1007/s12201-008-1003-8.

Clark, K. A., E. L. Krueger, and D. A. Vanden Bout. 2014. "Direct Measurement of Energy Migration in Supramolecular Carbocyanine Dye Nanotubes." *The Journal of Physical Chemistry Letters* no. 5 (13):2274-2282. doi: 10.1021/jz500634f.

Dahne, L. 1995. "Self-organization of polymethine dyes in thin solid layers." *Journal of the American Chemical Society* no. 117 (51):12855-12860.

Daniel, M. C., and D. Astruc. 2003. "Gold Nanoparticles: Assembly, Supramolecular Chemistry, Quantum-Size-Related Properties, and Applications toward Biology, Catalysis, and Nanotechnology." *Chemical Reviews* no. 104 (1):293-346. doi: 10.1021/cr030698+.

D'Avino, G., F. Terenziani, and A. Painelli. 2006. "Aggregates of Quadrupolar Dyes: Giant Two-Photon Absorption from Biexciton States." *The Journal of Physical Chemistry B* no. 110 (51):25590-25592. doi: 10.1021/jp0673361.

Davydov, A. S. 1963. "Theory of Molecular Excitons." *American Journal of Physics* no. 31 (3):220-220. doi: 10.1119/1.1969397.

Demchenko, A. 2013. *Nanoparticles and Nanocomposites for Fluorescence Sensing and Imaging.* Vol. 1.

Demchenko, A. P. 2010. "The concept of lambda-ratiometry in fluorescence sensing and imaging." *Journal of Fluorescence* no. 20 (5):1099-128. doi: 10.1007/s10895-010-0644-y.

Devaraj, P., P. Kumari, C. Aarti, and A. Renganathan. 2013. "Synthesis and Characterization of Silver Nanoparticles Using Cannonball Leaves and Their Cytotoxic Activity against MCF-7 Cell Line." *Journal of Nanotechnology* no. 2013:5. doi: 10.1155/2013/598328.

Dulkeith, E., A. C. Morteani, T. Niedereichholz, T. A. Klar, J. Feldmann, S. A. Levi, F. C. J. M. van Veggel, D. N. Reinhoudt, M. Möller, and D. I. Gittins. 2002. "Fluorescence Quenching of Dye Molecules near Gold Nanoparticles: Radiative and Nonradiative Effects." *Physical Review Letters* no. 89 (20):203002.

Dulkeith, E., M. Ringler, T. A. Klar, J. Feldmann, A. Muñoz Javier, and W. J. Parak. 2005. "Gold Nanoparticles Quench Fluorescence by Phase Induced Radiative Rate Suppression." *Nano Letters* no. 5 (4):585-589. doi: 10.1021/nl0480969.

Egorov, V. V. 2009. "Theory of the J-band: From the Frenkel exciton to charge transfer." *Physics Procedia* no. 2 (2):223-326. doi: https://doi.org/10.1016/j.phpro.2009.07.014.

Egorov, V. V. 2017. "Nature of the optical band shapes in polymethine dyes and H-aggregates: dozy chaos and excitons. Comparison with dimers, H*- and J-aggregates." *Royal Society Open Science* no. 4 (5):160550-160550. doi: 10.1098/rsos.160550.

Egorov, V., and M. Alfimov. 2007. "Theory of the J-band: from the Frenkel exciton to charge transfer." *Physics-Uspekhi* no. 50 (10):985.

Fofang, N. T., T.-H. Park, O. Neumann, N. A. Mirin, P. Nordlander, and N. J. Halas. 2008. "Plexcitonic Nanoparticles: Plasmon−Exciton Coupling in Nanoshell−J-Aggregate Complexes." *Nano Letters* no. 8 (10):3481-3487. doi: 10.1021/nl8024278.

Gadde, S., E. K. Batchelor, and A. Kaifer. 2009. "Controlling the Formation of Cyanine Dye H- and J-Aggregates with Cucurbituril Hosts in the Presence of Anionic Polyelectrolytes" *Chemistry - A European Journal* no. 15 (24):6025-31.

Gadde, S., E. K. Batchelor, J. P. Weiss, Y. Ling, and A. E. Kaifer. 2008. "Control of H- and J-Aggregate Formation via Host−Guest Complexation using Cucurbituril Hosts." *Journal of the American Chemical Society* no. 130 (50):17114-17119. doi: 10.1021/ja807197c.

Gesztelyi, R., J. Zsuga, A. Kemeny-Beke, B. Varga, B. Juhasz, and A. Tosaki. 2012. "The Hill equation and the origin of quantitative pharmacology." *Archive for History of Exact Sciences* no. 66 (4):427-438. doi: 10.1007/s00407-012-0098-5.

Golla, N., M. Koduru, and D. P. R. Borelli. 2011. *Mushrooms (Agaricus bisporus) mediated biosynthesis of sliver nanoparticles, characterization and their antimicrobial activity.* Vol. 2.

Görner, H., A. K. Chibisov, and T. D. Slavnova. 2006. "Kinetics of J-Aggregation of Cyanine Dyes in the Presence of Gelatin." *The Journal of Physical Chemistry B* no. 110 (9):3917-3923. doi: 10.1021/jp055876c.

Hannah, Kristen C., and Bruce A. Armitage. 2004. "DNA-Templated Assembly of Helical Cyanine Dye Aggregates: A Supramolecular Chain Polymerization." *Accounts of Chemical Research* no. 37 (11):845-853. doi: 10.1021/ar030257c.

Haverkort, F., A. Stradomska, A. Vries, and J. Knoester. 2013. *Investigating the Structure of Aggregates of an Amphiphilic Cyanine Dye with Molecular Dynamics Simulations.* Vol. 117.

Haverkort, Frank, Anna Stradomska, Alex H. de Vries, and Jasper Knoester. 2014. "First-Principles Calculation of the Optical Properties of an Amphiphilic Cyanine Dye Aggregate." *The Journal of Physical Chemistry A* no. 118 (6):1012-1023. doi: 10.1021/jp4112487.

Hestand, N. J., and F. C. Spano. 2018. "Expanded Theory of H- and J-Molecular Aggregates: The Effects of Vibronic Coupling and Intermolecular Charge Transfer." *Chemical Reviews* no. 118 (15):7069-7163. doi: 10.1021/acs.chemrev.7b00581.

Higgins, D. A., P. J. Reid, and P. F. Barbara. 1996. "Structure and Exciton Dynamics in J-Aggregates Studied by Polarization-Dependent Near-Field Scanning Optical Microscopy." *The Journal of Physical Chemistry* no. 100 (4):1174-1180. doi: 10.1021/jp9518217.

Hranisavljevic, J., N. M. Dimitrijevic, G. A. Wurtz, and G. P. Wiederrecht. 2002. "Photoinduced Charge Separation Reactions of J-Aggregates Coated on Silver Nanoparticles." *Journal of the American Chemical Society* no. 124 (17):4536-4537. doi: 10.1021/ja012263e.

Jang, S. J., and B. Mennucci. 2018. "Delocalized excitons in natural light-harvesting complexes." *Reviews of Modern Physics* no. 90 (3):035003. doi: 10.1103/RevModPhys.90.035003.

Jeunieau, L., V. Alin, and J. B. Nagy. 1999. "Adsorption of Thiacyanine Dyes on Silver Halide Nanoparticles: Study of the Adsorption Site." *Langmuir* no. 16 (2):597-606. doi: 10.1021/la990114s.

Johnston, D. C. 2006. "Stretched exponential relaxation arising from a continuous sum of exponential decays." *Physical Review B* no. 74 (18):184430. doi: 10.1103/PhysRevB.74.184430.

Jones, R. M., T. S. Bergstedt, C. T. Buscher, D. McBranch, and D. Whitten. 2001. "Superquenching and Its Applications in J-Aggregated Cyanine Polymers." *Langmuir* no. 17 (9):2568-2571. doi: 10.1021/la0017451.

Jung, B., V. I. Vullev, and B. Anvari. 2014. "Revisiting Indocyanine Green: Effects of Serum and Physiological Temperature on Absorption and Fluorescence Characteristics." *IEEE Journal of Selected Topics in Quantum Electronics* no. 20 (2):149-157. doi: 10.1109/JSTQE.2013.2278674.

Kamalov, V. F., I. A. Struganova, and K. Yoshihara. 1996. "Temperature Dependent Radiative Lifetime of J-Aggregates." *The Journal of Physical Chemistry* no. 100 (21):8640-8644. doi: 10.1021/jp9522472.

Kasha, M. 1963. "Energy Transfer Mechanisms and the Molecular Exciton Model for Molecular Aggregates." *Radiation Research* no. 20 (1):55-70. doi: 10.2307/3571331.

Kirstein, S., and S. Daehne. 2006. "J-aggregates of amphiphilic cyanine dyes: Self-organization of artificial light harvesting complexes." *International Journal of Photoenergy* no. 2006:21. doi: 10.1155/ijp/2006/20363.

Kneipp, J., H. Kneipp, B. Wittig, and K. Kneipp. 2009. *Novel optical nanosensors for probing and imaging live cells*. Vol. 6.

Kneipp, J., H. Kneipp, W. L. Rice, and K. Kneipp. 2005. "Optical Probes for Biological Applications Based on Surface-Enhanced Raman Scattering from Indocyanine Green on Gold Nanoparticles." *Analytical Chemistry* no. 77 (8):2381-2385. doi: 10.1021/ac050109v.

Kneipp, K., H. Kneipp, and J. Kneipp. 2006. "Surface-enhanced Raman scattering in local optical fields of silver and gold nanoaggregates-from single-molecule Raman spectroscopy to ultrasensitive probing in live cells." *Acc Chem Res* no. 39 (7):443-50. doi: 10.1021/ar050107x.

Kodaka, M. 2004. "Requirements for generating sigmoidal time–course aggregation in nucleation-dependent polymerization model." *Biophysical Chemistry* no. 107 (3):243-253. doi: http://dx.doi.org/10.1016/j.bpc.2003.09.013.

Kometani, N., M. Tsubonishi, T. Fujita, K. Asami, and Y. Yonezawa. 2001. "Preparation and Optical Absorption Spectra of Dye-Coated Au, Ag, and Au/Ag Colloidal Nanoparticles in Aqueous Solutions and in Alternate Assemblies." *Langmuir* no. 17 (3):578-580. doi: 10.1021/la0013190.

Laban, B., I. Zeković, D. Vasić Anićijević, M. Marković, V. Vodnik, M. Luce, A. Cricenti, M. Dramićanin, and V. Vasić. 2016. "Mechanism of 3,3′-Disulfopropyl-5,5′-Dichlorothiacyanine Anion Interaction With Citrate-Capped Silver Nanoparticles: Adsorption and J-Aggregation." *The Journal of Physical Chemistry C* no. 120 (32):18066-18074. doi: 10.1021/acs.jpcc.6b05124.

Laban, B., V. Vodnik, A. Vujačić, S. Sovilj, A. Jokić, and V. Vasić. 2013. "Spectroscopic and fluorescence properties of silver-dye composite nanoparticles." *Russian Journal of Physical Chemistry A* no. 87 (13):2219-2224. doi: 10.1134/S0036024413130141.

Laban, B., V. Vodnik, and V. Vasić. 2015. Spectrophotometric observations of thiacyanine dye J-aggregation on citrate capped silver nanoparticles. In: *Nanospectroscopy* no. 1 (1). doi: https://doi.org/10.1515/nansp-2015-0004

Laban, B., V. Vodnik, M. Dramićanin, M. Novaković, N. Bibić, S. P. Sovilj, and V. M. Vasić. 2014. "Mechanism and Kinetics of J-Aggregation of Thiacyanine Dye in the Presence of Silver Nanoparticles." *The Journal of Physical Chemistry C* no. 118 (40):23393-23401. doi: 10.1021/jp507086g.

Lakowicz, J. R. 2007. *Principles of Fluorescence Spectroscopy*: Springer.

Lazarević Pašti, T., V. Anićijević, M. Baljozović, D. Vasić Anićijević, S. Gutić, V. Vasić, N. V. Skorodumova, and I. A. Pašti. 2018. "The impact of the structure of graphene-based materials on the removal of organophosphorus pesticides from water." *Environmental Science: Nano* no. 5 (6):1482-1494. doi: 10.1039/C8EN00171E.

Lazarević-Pašti, T. D., I. A. Pašti, B. Jokić, B. M. Babić, and V. M. Vasić. 2016. "Heteroatom-doped mesoporous carbons as efficient adsorbents for removal of dimethoate and omethoate from water." *RSC Advances* no. 6 (67):62128-62139. doi: 10.1039/C6RA06736K.

Lazarević-Pašti, T., V. Anićijević, M. Baljozović, D. Vasić Anićijević, S. Gutić, V. Vasić, N. V. Skorodumova, and I. A. Pašti. 2018. "The impact of the structure of graphene-based materials on the removal of organophosphorus pesticides from water." *Environmental Science: Nano* no. 5 (6):1482-1494. doi: 10.1039/C8EN00171E.

Lebedev, V. S., A. G. Vitukhnovsky, A. Yoshida, N. Kometani, and Y. Yonezawa. 2008. "Absorption properties of the composite silver/dye nanoparticles in colloidal solutions." *Colloids and Surfaces A: Physicochemical and Engineering Aspects* no. 326 (3):204-209. doi: 10.1016/j.colsurfa.2008.06.027.

Lebedev, V. S., and A. S. Medvedev. 2013a. "Optical properties of three-layer metal-organic nanoparticles with a molecular J-aggregate shell." *Quantum Electronics* no. 43 (11):1065-1077.

Lebedev, V. S., and A. S. Medvedev. 2013b. "Absorption and Scattering of Light by Hybrid Metal/J-Aggregate Nanoparticles: Plasmon–Exciton Coupling and Size Effects." *Journal of Russian Laser Research* no. 34 (4):303-322. doi: 10.1007/s10946-013-9356-5.

Lebedev, Vladimir S., A. S. Medvedev, D. N. Vasil'ev, D. A. Chubich, and Alexey G. Vitukhnovsky. 2010. "Optical properties of noble-metal nanoparticles coated with a dye J-aggregate monolayer." *Quantum Electronics* no. 40 (3):246-253. doi: 10.1070/qe2010v040n03 abeh014209.

Leroy, P., C. Tournassat, and M. Bizi. 2011. "Influence of surface conductivity on the apparent zeta potential of TiO2 nanoparticles."

Journal of Colloid and Interface Science no. 356 (2):442-53. doi: 10.1016/j.jcis.2011.01.016.

Leroy, P., N. Devau, A. Revil, and M. Bizi. 2013. "Influence of surface conductivity on the apparent zeta potential of amorphous silica nanoparticles." *J Colloid Interface Sci* no. 410:81-93. doi: 10.1016/j.jcis.2013.08.012.

Lewis, E. S.. 1974. *Investigation of Rates and Mechanisms of Reactions, Part I.* 3rd ed. New York: Wiley-Interscience: .

Lim, I. I. S., F. Goroleski, D. Mott, N. Kariuki, W. Ip, J. Luo, and C. J. Zhong. 2006a. "Adsorption of Cyanine Dyes on Gold Nanoparticles and Formation of J-Aggregates in the Nanoparticle Assembly." *The Journal of Physical Chemistry B* no. 110 (13):6673-6682. doi: 10.1021/jp057584h.

Lim, I. I. S., F. Goroleski, D. Mott, N. Kariuki, W. Ip, J. Luo, and C.-J. Zhong. 2006b. "Adsorption of Cyanine Dyes on Gold Nanoparticles and Formation of J-Aggregates in the Nanoparticle Assembly." *The Journal of Physical Chemistry B* no. 110 (13):6673-6682. doi: 10.1021/jp057584h.

Lin, H., R. Camacho, Y. Tian, T. E. Kaiser, F. Würthner, and I. G. Scheblykin. 2010. "Collective Fluorescence Blinking in Linear J-Aggregates Assisted by Long-Distance Exciton Migration." *Nano Letters* no. 10 (2):620-626. doi: 10.1021/nl9036559.

Link, S., and M. A. El-Sayed. 1999. "Spectral Properties and Relaxation Dynamics of Surface Plasmon Electronic Oscillations in Gold and Silver Nanodots and Nanorods." *The Journal of Physical Chemistry B* no. 103 (40):8410-8426. doi: 10.1021/jp9917648.

Liu, F., B. Sanyasi Rao, and J. M. Nunzi. 2011. "A dye functionalized silver–silica core–shell nanoparticle organic light emitting diode." *Organic Electronics* no. 12 (7):1279-1284. doi: 10.1016/j.orgel.2011. 04.013.

Majhi, D., S. K. Das, P. K. Sahu, S. M. Pratik, A. Kumar, and M. Sarkar. 2014. "Probing the aggregation behavior of 4-aminophthalimide and 4-(N,N-dimethyl) amino-N-methylphthalimide: a combined photophysical, crystallographic, microscopic and theoretical (DFT) study." *Physical Chemistry Chemical Physics* no. 16 (34):18349-18359. doi: 10.1039/C4CP01912A.

Maslovara, S., D. Vasić Anićijević, S. Brković, J. Georgijević, G. Tasić, and M. Marčeta Kaninski. 2019. "Experimental and DFT study of CoCuMo ternary ionic activator for alkaline HER on Ni cathode." *Journal of Electroanalytical Chemistry* no. 839:224-230. doi: https://doi.org/10.1016/j.jelechem.2019.03.044.

Merdasa, A., A. Jimenez, R. Camacho, M. Meyer, F. Wurthner, and I. G. Scheblykin. 2014. "Single Levy states-disorder induced energy funnels in molecular aggregates." *Nano Letters* no. 14 (12):6774-81. doi: 10.1021/nl5021188.

Milović, M. D., D. D. Vasić Anicijević, D. Jugović, V. J. Anicijević, L. Veselinović, M. Mitrić, and D. Uskoković. 2019. "On the presence of antisite defect in monoclinic Li2FeSiO4 - A combined X-Ray diffraction and DFT study." *Solid State Sciences* no. 87:81-86. doi: 10.1016/j.solidstatesciences.2018.11.008.

Mirković, M. M., T. D. Lazarević Pašti, A. M. Došen, M. Ž Čebela, A. A. Rosić, B. Z. Matović, and B. M. Babić. 2016. "Adsorption of malathion on mesoporous monetite obtained by mechanochemical treatment of brushite." *RSC Advances* no. 6 (15):12219-12225. doi: 10.1039/C5RA27554G.

Mishra, A., R. K. Behera, P. K. Behera, I. K. Mishra, and G. B. Behera. 2000. "Cyanines during the 1990s: A Review." *Chemical Reviews* no. 100 (6):1973-2012. doi: 10.1021/cr990402t.

Moerner, W. E., and L. Kador. 1989. "Optical detection and spectroscopy of single molecules in a solid." *Physical Review Letters* no. 62 (21):2535-2538. doi: 10.1103/PhysRevLett.62.2535.

Momić, T., T. Lazarević Pašti, U. Bogdanović, V. Vodnik, A. Mraković, Z. Rakočević, V. B. Pavlović, and V. Vasić. 2016. "Adsorption of Organophosphate Pesticide Dimethoate on Gold Nanospheres and Nanorods." *Journal of Nanomaterials* no. 2016:11. doi: 10.1155/2016/8910271.

Munson, P. J. 1984. "Ligand Binding Data Analysis: Theoretical and Practical Aspects." In *Principles and Methods in Receptor Binding*, edited by F. Cattabeni and S. Nicosia, 1-12. Springer US.

Neves, T. B. V., and G. F. S. Andrade. 2015. "SERS Characterization of the Indocyanine-Type Dye IR-820 on Gold and Silver Nanoparticles in the Near Infrared." *Journal of Spectroscopy* no. 2015:9. doi: 10.1155/2015/805649.

Oba, T., and H. Tamiaki. 1998. "Molecular Requirement of Chlorosomal Chlorophylls. Self-Organization of a Chlorophyll Derivative Possessing a Hydroxyl Group at Ring II." *Photochemistry and Photobiology* no. 67 (3):295-303. doi: 10.1111/j.1751-1097.1998. tb05202.x.

Orrit, M., and J. Bernard. 1990. "Single pentacene molecules detected by fluorescence excitation in a p-terphenyl crystal." *Physical Review Letters* no. 65 (21):2716-2719. doi: 10.1103/PhysRevLett.65.2716.

Osedach, T. P., A. Iacchetti, R. R. Lunt, T. L. Andrew, P. R. Brown, G. M. Akselrod, and V. Bulović. 2012. "Near-infrared photodetector consisting of J-aggregating cyanine dye and metal oxide thin films." *Applied Physics Letters* no. 101 (11):113303. doi: 10.1063/1. 4752434.

Pasternack, R. F., C. Fleming, S. Herring, P. J. Collings, J. dePaula, G. DeCastro, and E. J. Gibbs. 2000. "Aggregation Kinetics of Extended Porphyrin and Cyanine Dye Assemblies." *Biophysical journal* no. 79 (1):550-560.

Pastore, M., and F. De Angelis. 2010. "Aggregation of Organic Dyes on TiO2 in Dye-Sensitized Solar Cells Models: An ab Initio Investigation." *ACS Nano* no. 4 (1):556-562. doi: 10.1021/nn 901518s.

Patra, A., C. G. Chandaluri, and T. P. Radhakrishnan. 2012. "Optical materials based on molecular nanoparticles." *Nanoscale* no. 4 (2):343-359. doi: 10.1039/C1NR11313E.

Pisoni, D. S., L. Todeschini, A. C. A. Borges, C. L. Petzhold, F. S. Rodembusch, and L. F. Campo. 2014. "Symmetrical and Asymmetrical Cyanine Dyes. Synthesis, Spectral Properties, and

BSA Association Study." *The Journal of Organic Chemistry* no. 79 (12):5511-5520. doi: 10.1021/jo500657s.

Prokhorov, V. V., S. I. Pozin, D. A. Lypenko, O. M. Perelygina, E. I. Mal'tsev, and A. V. Vannikov. 2012. "Molecular arrangements in two-dimensional J-aggregate monolayers of cyanine dyes." *Macroheterocycles* no. 5 (4-5):371-376.

Ralević, U., G. Isić, D. Vasić Anićijević, B. Laban, U. Bogdanović, V. M. Lazović, V. Vodnik, and R. Gajić. 2018. "Nanospectroscopy of thiacyanine dye molecules adsorbed on silver nanoparticle clusters." *Applied Surface Science* no. 434:540-548. doi: https://doi.org/10.1016/j.apsusc.2017.10.148.

Ravindran, A., P. Chandran, and S. S. Khan. 2013. "Biofunctionalized silver nanoparticles: Advances and prospects." *Colloids and Surfaces B: Biointerfaces* no. 105:342-352. doi: https://doi.org/10.1016/j.colsurfb.2012.07.036.

Reinheimer, J. D. 1965. "Investigation of rates and mechanisms of reactions. Parts 1 and 2 (Friess, S. L.; Weis, E. S.; Weissberger, Arnold; eds.)." *Journal of Chemical Education* no. 42 (2):A148. doi: 10.1021/ed042pA148.2.

Saccone, D., S. Galliano, N. Barbero, P. Quagliotto, G. Viscardi, and C. Barolo. 2016. "Polymethine Dyes in Hybrid Photovoltaics: Structure–Properties Relationships." *European Journal of Organic Chemistry* no. 2016 (13):2244-2259. doi: 10.1002/ejoc.201501598.

Saleh, T. A. 2014. "Spectroscopy: Between Modeling, Simulation and Practical Investigation." *Spectral Analysis Review* no. 2014 (01):1-2. doi: 10.4236/sar.2014.21001.

Scatchard, G. 1949. "The Attractions Of Proteins For Small Molecules And Ions." *Annals of the New York Academy of Sciences* no. 51 (4):660-672. doi: 10.1111/j.1749-6632.1949.tb27297.x.

Scheblykin, I. G., M. A. Drobizhev, O. P. Varnavsky, M. Van der Auweraer, and A. G. Vitukhnovsky. 1996. "Reorientation of transition dipoles during exciton relaxation in J-aggregates probed by flourescence anisotropy." *Chemical Physics Letters* no. 261 (1):181-190. doi: https://doi.org/10.1016/0009-2614(96)00946-3.

Scheblykin, I. G., O. Yu Sliusarenko, L. S. Lepnev, A. G. Vitukhnovsky, and M. Van der Auweraer. 2001. "Excitons in Molecular Aggregates of 3,3'-Bis-[3-sulfopropyl]-5,5'-dichloro-9- ethylthiacarbocyanine (THIATS): Temperature Dependent Properties." *The Journal of Physical Chemistry B* no. 105 (20):4636-4646. doi: 10.1021/jp004294m.

Schwartzberg, A. M., C. D. Grant, A. Wolcott, C. E. Talley, T. R. Huser, R. Bogomolni, and J. Z. Zhang. 2004. "Unique Gold Nanoparticle Aggregates as a Highly Active Surface-Enhanced Raman Scattering Substrate." *The Journal of Physical Chemistry B* no. 108 (50):19191-19197. doi: 10.1021/jp048430p.

Shindy, H. A. 2017. "Fundamentals in the chemistry of cyanine dyes: A review." *Dyes and Pigments* no. 145:505-513. doi: https://doi.org/10.1016/j.dyepig.2017.06.029.

Siwach, O. P., and P. Sen. 2009. "Fluorescence properties of Ag nanoparticles in water, methanol and hexane." *Journal of Luminescence* no. 129 (1):6-11. doi: 10.1016/j.jlumin.2008.07.010.

Slavnova, T. D., A. K. Chibisov, and H. Görner. 2005. "Kinetics of Salt-Induced J-aggregation of Cyanine Dyes." *The Journal of Physical Chemistry A* no. 109 (21):4758-4765. doi: 10.1021/jp058014k.

Sorokin, A. V., A. A. Zabolotskii, N. V. Pereverzev, S. L. Yefimova, Y. V. Malyukin, and A. I. Plekhanov. 2014. "Plasmon Controlled Exciton Fluorescence of Molecular Aggregates." *The Journal of Physical Chemistry C* no. 118 (14):7599-7605. doi: 10.1021/jp412798u.

Struganova, I. A., H. Lim, and S. A. Morgan. 2002. "The Influence of Inorganic Salts and Bases on the Formation of the J-band in the Absorption and Fluorescence Spectra of the Diluted Aqueous Solutions of TDBC." *The Journal of Physical Chemistry B* no. 106 (42):11047-11050. doi: 10.1021/jp013511w.

Struganova, I. A., M. Hazell, J. Gaitor, D. McNally-Carr, and S. Zivanovic. 2003. "Influence of Inorganic Salts and Bases on the J-Band in the Absorption Spectra of Water Solutions of 1,1'-Diethyl-2,2'-cyanine Iodide." *The Journal of Physical Chemistry A* no. 107 (15):2650-2656. doi: 10.1021/jp0223004.

Tani, T. 2011. *Photographic Science: Advances in Nanoparticles, J-Aggregates, Dye Sensitization, and Organic Devices.* . Oxford: Oxford University Press, Oxford Scholarship Online, 2012.

Tatikolov, A. S. 2012. "Polymethine dyes as spectral-fluorescent probes for biomacromolecules." *Journal of Photochemistry and Photobiology C: Photochemistry Reviews* no. 13 (1):55-90. doi: https://doi.org/10.1016/j.jphotochemrev.2011.11.001.

Tillmann, W., and H. Samha. 2004. "J-Aggregates of Cyanine Dyes in Aqueous Solution of Polymers: A Quantitative Study." *American Journal of Undergraduate Research* no. 3 (3):1-6.

Tomioka, A., S. Kinoshita, and A. Fujimoto. 2007. "Molecular Ordering in Self-Organized Dye Particles." *Molecular Crystals and Liquid Crystals* no. 471 (1):69-80. doi: 10.1080/15421400701545296.

Tran, Q. H., V. Q. Nguyen, and A. T. Le. 2013. "Silver nanoparticles: synthesis, properties, toxicology, applications and perspectives." *Advances in Natural Sciences: Nanoscience and Nanotechnology* no. 4 (3):033001. doi: 10.1088/2043-6262/4/3/033001.

Vasić Anićijević, D., I. Perović, S. Maslovara, S. Brković, D. Žugić, Z. Laušević, and M. Marčeta Kaninski. 2016. "Ab initio study of graphene interaction with O2, O and O-." *Macedonian Journal of Chemistry and Chemical Engineering* no. 35 (2):271-274.

Vasić Anićijević, D., V. Nikolić, M. Marčeta Kaninski, and I. Pašti. 2015. "Structure, chemisorption properties and electrocatalysis by Pd3Au overlayers on tungsten carbide – A DFT study." *International Journal of Hydrogen Energy* no. 40 (18):6085-6096. doi: https://doi.org/10.1016/j.ijhydene.2015.03.083.

Vodnik, V., and J. Nedeljković. 2000. "Influence of Negative Charge on the Optical Properties of a Silver Sol." *Journal of the Serbian Chemical Society* no. 65 (3):195 - 200.

Voet, D., and J. Voet. 1993. *Biochemistry.* New York: John Willey & Sons.

Voznyak, D. A., and A. K. Chibisov. 2008. "Kinetic models of J-aggregation of polymethine dyes." *Nanotechnologies in Russia* no. 3 (9-10):543-550. doi: 10.1134/S1995078008090024.

Vujačić, A., V. Vasić, M. Dramićanin, S. P. Sovilj, N. Bibić, J. Hranisavljevic, and G. P. Wiederrecht. 2012. "Kinetics of J-Aggregate Formation on the Surface of Au Nanoparticle Colloids." *The Journal of Physical Chemistry C* no. 116 (7):4655-4661. doi: 10.1021/jp210549u.

Vujačić, A., V. Vasić, M. Dramićanin, S. P. Sovilj, N. Bibić, S. Milonjić, and V. Vodnik. 2013. "Fluorescence Quenching of 5,5'-Disulfopropyl-3,3'-dichlorothiacyanine Dye Adsorbed on Gold Nanoparticles." *The Journal of Physical Chemistry C* no. 117 (13):6567-6577. doi: 10.1021/jp311015w.

Vujačić, A., V. Vodnik, S. P. Sovilj, M. Dramićanin, N. Bibić, S. Milonjić, and V. Vasić. 2013. "Adsorption and fluorescence quenching of 5,5[prime or minute]-disulfopropyl-3,3[prime or minute]-dichlorothiacyanine dye on gold nanoparticles." *New Journal of Chemistry* no. 37 (3):743-751. doi: 10.1039/C2NJ40865A.

Wang, Yi., and X. Zhan. 2016. "Layer-by-Layer Processed Organic Solar Cells." *Advanced Energy Materials* no. 6 (17):1600414. doi: 10.1002/aenm.201600414.

West, W., and S. Pearce. 1965. "The Dimeric State of Cyanine Dyes." *The Journal of Physical Chemistry* no. 69 (6):1894-1903. doi: 10.1021/j100890a019.

Wiederrecht, G. P., G. A. Wurtz, and J. Hranisavljevic. 2004. "Coherent Coupling of Molecular Excitons to Electronic Polarizations of Noble Metal Nanoparticles." *Nano Letters* no. 4 (11):2121-2125. doi: 10.1021/nl0488228.

Wulandari, P., T. Nagahiro, N. Fukada, Y. Kimura, M. Niwano, and K. Tamada. 2015. "Characterization of citrates on gold and silver nanoparticles." *Journal of Colloid and Interface Science* no. 438:244-248. doi: https://doi.org/10.1016/j.jcis.2014.09.078.

Xu, R. 2008. "Progress in nanoparticles characterization: Sizing and zeta potential measurement." *Particuology* no. 6 (2):112-115. doi: https://doi.org/10.1016/j.partic.2007.12.002.

Yao, H., K. Domoto, T. Isohashi, and K. Kimura. 2005. "In Situ Detection of Birefringent Mesoscopic H and J Aggregates of Thiacarbocyanine

Dye in Solution." *Langmuir* no. 21 (3):1067-1073. doi: 10.1021/la0479004.

Yao, H., T. Isohashi, and K. Kimura. 2004. "Large birefringence of single J aggregate nanosheets of thiacyanine dye in solution." *Chemical Physics Letters* no. 396 (4):316-322. doi: https://doi.org/10.1016/j.cplett.2004.08.045.

Yao, H., T. Isohashi, and K. Kimura. 2007. "Electrolyte-Induced Mesoscopic Aggregation of Thiacarbocyanine Dye in Aqueous Solution: Counterion Size Specificity." *The Journal of Physical Chemistry B* no. 111 (25):7176-7183. doi: 10.1021/jp070520h.

Yoshida, A., N. Kometani, and Y. Yonezawa. 2008. "Silver:dye composite nanoparticles as a building unit of molecular architecture." *Colloids and Surfaces A: Physicochemical and Engineering Aspects* no. 313–314 (0):581-584. doi: 10.1016/j.colsurfa.2007.04.165.

Yoshida, A., N. Uchida, and N. Kometani. 2009. "Synthesis and Spectroscopic Studies of Composite Gold Nanorods with a Double-Shell Structure Composed of Spacer and Cyanine Dye J-Aggregate Layers." *Langmuir* no. 25 (19):11802-11807. doi: 10.1021/la901431r.

Yoshida, A., Y. Yonezawa, and N. Kometani. 2009. "Tuning of the Spectroscopic Properties of Composite Nanoparticles by the Insertion of a Spacer Layer: Effect of Exciton−Plasmon Coupling." *Langmuir* no. 25 (12):6683-6689. doi: 10.1021/la900169e.

Zhang, A., Y. Fang, and H. Shao. 2006. "Studies of quenching and enhancement of fluorescence of methyl orange adsorbed on silver colloid." *Journal of Colloid and Interface Science* no. 298 (2):769-772. doi: 10.1016/j.jcis.2006.01.014.

Zhang, H., Q Wu, and M. Y. Berezin. 2015. "Fluorescence anisotropy (polarization): from drug screening to precision medicine." *Expert Opinion on Drug Discovery* no. 10 (11):1145-1161.

Zhang, X. F., Z. G. Liu, W. Shen, and S. Gurunathan. 2016. "Silver Nanoparticles: Synthesis, Characterization, Properties, Applications, and Therapeutic Approaches." *International Journal of Molecular Sciences* no. 17 (9):1534.

Zhong, Z., J. Luo, T. P. Ang, J. Highfield, J. Lin, and A. Gedanken. 2004. "Controlled Organization of Au Colloids into Linear Assemblies." *The Journal of Physical Chemistry B* no. 108 (47):18119-18123. doi: 10.1021/jp047683f.

In: Cyanine Dyes
Editor: Douglas Zimmerman

ISBN: 978-1-53616-239-4
© 2019 Nova Science Publishers, Inc.

Chapter 4

CYANINE DYE AGGREGATION WITHIN IONIC LIQUID AND DEEP EUTECTIC SOLVENT BASED SYSTEMS: A REVIEW

Bhawna[*], Divya Dhingra[*] and Siddharth Pandey[†]
Department of Chemistry, Indian Institute of Technology Delhi,
Delhi, India

ABSTRACT

The self-aggregates of cyanine dyes owning interesting photo-physical properties have evoked great curiosity due to their remarkable technological applications in the fields, such as photography, sensors, photoconductors, medicine, and nanotechnology, among others. The highly ordered aggregates of various structures and morphologies of cyanine dyes are governed by the solvent-dye interactions. So far, while aqueous systems have been reported to be the most favorable media for cyanine dye aggregation, little has been known about dye aggregation in non-aqueous media. Ionic liquids (ILs) and deep eutectic solvents (DESs), due to their unique physicochemical properties, have shown

---

[*] These authors contributed equally to this work.
[†] Corresponding Author's E-mail: sipandey@chemistry.iitd.ac.in.

immense potential as non-aqueous media for the molecular aggregation, though the reports are few and scarce. Recently, interesting studies have surfaced showing the dependence of cyanine dye aggregation on the identity of ILs and DESs. Thus, the unique properties displayed by ILs and DESs together with the unusual aggregation behavior of cyanine dyes make this field of research more attractive. With the aim to encourage further developments, this chapter focuses on exploring the interactions and aggregation behavior of cyanine dyes within various IL and DES-based systems.

Keywords: cyanine dye, ionic liquids, deep eutectic solvents, H- and J-aggregates

1. INTRODUCTION

1.1. Overview

Solvent occupies a strategic place in chemistry due to the diversified roles adhered to it. It is an essential component deciding the efficiency, stability, and kinetics of a chemical reaction. Hence, solvent selection becomes more important for the success of any particular chemical reaction [1–5]. Most of the solvents employed in academia and industries are traditional volatile organic compounds (VOCs). However, they have a detrimental impact on the global environment and human health due to their high flammability, toxicity, and volatility [6–11]. Therefore, it is highly essential to find an environment-friendly alternative for harmful VOCs. Researchers in the area of chemical sciences are making a conscious approach towards the development of environmentally-benign/green solvents [12–18]. In this regard, ionic liquids (ILs) and deep eutectic solvents (DESs) have garnered substantial attention as a class of green solvents because of their intriguing physical and chemical properties. These solvents are regarded as viable alternatives for VOCs, due to their easy availability, non-toxicity, biodegradability, recyclability, and low flammability. In this chapter, we explore the applicability of ILs and DESs as alternate solvent media for the cyanine dye aggregation phenomena.

1.2. Ionic Liquids

The term IL describes a wide class of low melting semi-organic/inorganic salts which exist in the liquid state at the ambient conditions. They are mostly composed of a bulky organic cation and an inorganic or organic polyatomic anion. History of ILs is endless and several stories related to their discovery and applications exist in literature. In 1888, Gabriel reported the first ever IL which was based on ammonium-cation with nitrate as an anion (ethanolammonium nitrate, EOAN) [19]. Later, Harley and Weir at Rice University led the first major studies on room temperature molten salts, wherein, the electrochemical properties of fused mixtures of ethylpyridinium bromide and aluminium chlorides were reported. Moreover, findings on 1-ethyl-3-methylimidazolium (emim) chloroaluminates established the existence of alkyl-based imidazolium molten salts [20, 21]. Over the last two decades, science and technology are benefitted immensely by the existence of a large number of ILs having inherently unique physicochemical characteristics such as low volatility, thermal and oxidative stability, tunable viscosity, broad electrochemical window, low flammability, high heat capacity, and among many others [22–24]. Literature suggests that such distinctive features of ILs are a result of coulombic, hydrogen bonding and van der waals interactions between their ions [25]. Interestingly, a wide range of ILs can be synthesized by using a large pool of cations and anions (Figure 1), simultaneously their physicochemical properties can also be tuned on modifying either cation or anion.

Moreover, ILs can be broadly categorized into four different segments based on their cationic components, a) di-substituted-alkylimidazolium-, b) ammonium-, c) phosphonium-, and d) N-alkyl-pyridinium-based ILs [26]. Owing to the ease of synthesis, low viscosity, and stability towards oxidation-reduction processes, the molten salts having imidazolium-based cationic moieties are the most studied [27–31]. Applications of imidazolium-based ILs include their potential usage as a lubricant in oil, supporting media in CNT formation, and in catalysis for enhancing chemo-selectivity, yield as well as reaction kinetics [32, 33]. Compared to the imidazolium-based cationic counterparts, ILs with pyridinium based-cationic components are more novel and therefore less popular as the investigation of their physicochemical properties and applications in various fields is yet to be fully ascertained [34–40]. On the contrary, remarkable thermal stability (~400 °C) of phosphonium-based ILs make them quite popular as a viable solvent media for chemical reactions carried out at a significantly higher temperature and also in CO_2 capture [41–47]. The ammonium-based ILs and their potential applications are less fancied and reported as compared to their counterparts [48–51].

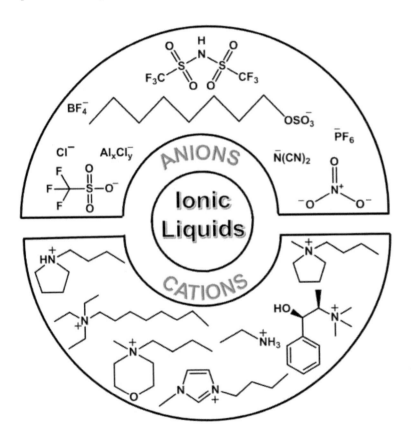

Figure 1. Different cationic and anionic components of ILs.

It is obligatory to mention the significance of anionic moiety in ILs as they play an essential role in shaping their physicochemical properties.

Factors such as the size of anion and asymmetric charge distribution govern the melting point of ILs, e.g., ILs containing halide ions have a higher melting point than those having other anions [52]. Moreover, the size of an anion also dictates the thermal stability of ILs, e.g., the highest decomposition temperature is shown for ILs having the large anion $[Tf_2N]^-$ [53]. Also, the thermal stability of ILs increases with increase in anionic charge density [54, 55]. The miscibility of ILs with other solvents, which is an important factor governing their wide applications, is largely influenced by the identity of their cation and anion [56]. For example, imidazolium-based ILs when coupled with anions such as nitrate, halide, ethanoate, and trifluoroacetate result in their complete miscibility with water [57]. Further, the miscibility of water with IL decreases with an increase in the alkyl chain length of the cation.

The exciting physical properties of ILs, such as density and viscosity, can be drastically altered by the size of the cationic ring, symmetry of ions and the length of the alkyl chain in the cation. In addition, the nature of hydrogen bonding and van der waals interactions play a decisive role in defining these physical properties. Literature reports that the density of several ILs show a linear dependency on the symmetry of cations and inverse with the size of the organic cations present in them [58]. ILs having longer and fluorinated alkyl chain have higher viscosities due to the presence of stronger van der waals interactions [59, 60]. Moreover, the viscosity of ILs may also be affected by the electrostatic interactions existing between the ions and the dispersion of electronic charge at the anion. Therefore, fine tuning of the above physical properties can be achieved by careful selection of anions and cations [61–68].

In recent years, researchers across the world have shown significant interest in the development of task-specific ionic liquids (TSILs) [69]. They are synthesized by incorporating some acidic or basic functionalities into either cation or anion moieties of the ILs, e.g., $-COOH$, $-SO_3H$ groups in the cation segment in combination with HSO_4^-, $H_2PO_4^-$, $H(HF)_{2,3}^-$, like anions leads to the formation of ILs with acidic properties. Moreover, the combination of cations containing $-NH_2$, pyrazine, pyrimidine with Cl^-

, CH_3COO^-, and cyanamide anions produce basic ILs [70–77]. They have found potential applications in various field such as nanomaterials synthesis, catalysis, and phase separation process including many others. Davi and co-workers predominantly reported the use of TSILs in the extraction of heavy metal ions e.g., Hg^{2+}, Cd^{2+} present in aqueous media [78].

The fascinating properties of these environmentally-benign liquids are becoming progressively popular within academia and industrial research [79, 80]. The realization of remarkable properties and full potential of ILs have led to their rapid development and exploration in numerous fields (Figure 2) which is reflected by an upsurge in the number of publications on these green solvents [81–83].

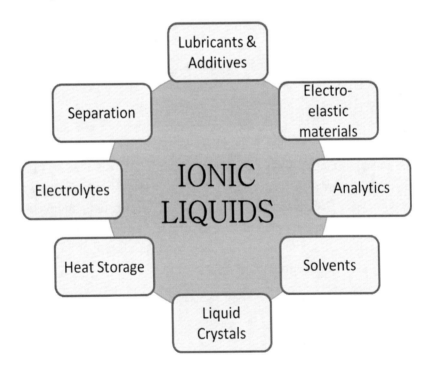

Figure 2. Various applications of ILs in numerous fields.

ILs have useful applications in many fields such as organic synthesis, electrochemistry, liquid phase extraction, catalysis, and polymerization processes. Nonetheless, these environmentally-benign solvents have emerged as a potential replacement for commonly used hazardous organic solvents [84, 85]. However, a lot of research is yet to be pursued which may contribute to more sustainable industrial applications of this green medium in the near future [86, 87].

1.3. Deep Eutectic Solvents (DESs)

DESs are new generation green solvents composed of two or more benign components which associate to form a eutectic mixture characterized by melting point significantly lower than the individual components. They are generally obtained by the complexation between a hydrogen bond acceptor (HBA) and a hydrogen bond donor (HBD) through hydrogen bonding interactions. These interactions result in charge delocalization between the two components which accounts for the large depression in the melting point of these solvents [88–91]. It has been proposed that the components with greater hydrogen-bonding ability exhibit larger depression in the freezing point. A broad range of DESs can be prepared by changing the HBA, almost infinitely, in combination with a very large choice of HBDs. This possibility of structure modification has earned them the title of designer solvents [92–94].

Abbott and co-workers, for the very first time, introduced DESs as versatile alternatives for conventional organic solvents and ILs. In 2003, they made pioneering development by preparing a eutectic solvent from urea and choline chloride (ChCl), which are environmentally-benign materials with high melting points. To signify the large depression in freezing point of the obtained mixture, the authors added prefix 'deep' to the eutectic solvent, thus, coining the term "deep eutectic solvents". This led to a growing interest in search of new HBA and HBD [95, 96]. Thereafter, a whole new range of DESs were discovered broadening the scope of their applications. In exemplary of this, Choi and co-workers reported 30 viscous eutectic mixtures, referred as 'natural deep eutectic solvents (NADES)', by combining ChCl with different HBDs such as organic acids, amino acids, and sugars [97]. The significantly low toxicity of these solvents inspired other researchers to further explore them. Most studied DESs are hydrophilic in nature showing miscibility with aqueous systems. On the contrary, Kroon and co-workers reported a new class of water-immiscible DES, predominantly known as hydrophobic DES [98]. Over the years, DESs have gained unsurpassed attention of both industry and scientific community and the number of publications dedicated to them is also escalating, further validating the fascination of this media.

Abbott and co-workers generalised these novel solvents as:

$$A^+B^-nY \tag{1}$$

Where A^+ is any quaternary ammonium, phosphonium or sulfonium cation, B is a Lewis base, generally a halide anion and n is the number of Y molecules that interact with the anion [99, 100]. They further categorized DESs on the basis of the nature of complexing components as shown in Figure 3.

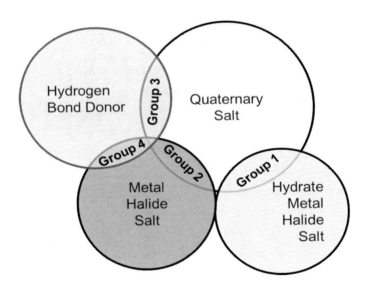

Figure 3. Different categories of DESs.

1.3.1. Group I DESs

They are obtained by mixing quaternary ammonium salts with well-known metal halides such as $FeCl_3$, $ZnCl_2$, $SnCl_2$, and many more. However, their applications are limited due to the high price of metal halides. Some of the commonly used group I DESs include the chloroaluminate/imidazolium salts and imidazolium/$FeCl_2$ mixtures [101].

1.3.2. Group II DESs

These are the mixtures of quaternary ammonium salts and the relatively cheaper hydrated metal halides, instead of simple metal halides. Low cost of the hydrated salts along with their insensitivity towards air and water makes them viable candidates for industrial applications [102].

1.3.3. Group III DESs

This class of DESs has created a lot of interest among researchers since they are prepared from non-toxic, inexpensive and ecofriendly materials. Figure 4 summarizes different quaternary ammonium salts that are widely used in combination with various HBDs to form DESs. Most of the DESs reported in literature belong to this class and are commonly based on choline chloride (ChCl) as the quaternary ammonium salt. ChCl (2-hydroxyethyltrimethylammonium chloride), which is derived from biomass, is the most talked quaternary ammonium salt due to the non-toxicity, biodegradability, easy extraction/synthesis, and relatively low cost of the salt compared to other inorganic salts [103].

1.3.4. Group IV DESs

These are the mixtures of transition metal halide with a variety of HBDs. Though recent reports on the formation of eutectics using $ZnCl_2$ with urea, acetamide, ethylene glycol, and 1,6-hexanediol have surfaced, the research on group IV DESs is still in its infancy [104].

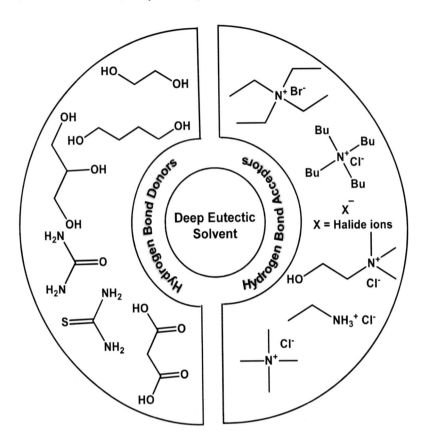

Figure 4. Various combination of HBDs and HBAs in DESs.

DESs are also known as analogues of ILs, due to their notable IL-like physico-chemical properties like good thermal stability, high conductivity, negligible vapor pressure, wide electrochemical window, non-flammability, etc. Similar to ILs, they can also be chemically tailored to a certain set of properties by varying HBA or HBD. As, the physical properties of DES such as density and viscosity largely depends on the starting components, their molar ratio, and the molecular organization of DES [105]. Thus, the possibility to customize task-specific DES has stimulated much of the excitement among researchers. Moreover, they possess several advantages over conventional ILs, such as their ease of preparation, easy storage, and high purity. Compared to ILs, DESs can be easily prepared from lower cost, readily available, and biodegradable components by simply mixing and heating them with constant stirring till a homogenous clear liquid is obtained [106]. This synthesis is 100% atom economic, easy to handle, and requires no further purification. These advantages further provide ample opportunities for researchers to further quantify and qualify the greenness of these solvents. On the contrary to ILs, recycling of the DESs can be achieved by using various techniques such as pressure-driven membrane processes, nanofiltration, reverse osmosis, and pervaporation [107]. Also, DESs have potential to dissolve a broad variety of materials, such as proteins, salts, drugs, surfactants, polysaccharides, amino acids, and sugars due to their capability of accepting or donating protons. Owing to these remarkable advantages, DESs prove to be economically and technically the most viable alternative for traditional organic solvents [108, 109].

They are widely acknowledged as a new class of promising green solvents for the 21st century. With the great advantage of low ecological footprint and attractive price, DESs have experienced rapid development and exploration in numerous fields, such as electrochemistry where they are being used for electrodeposition of metals and as electrolytes for dye-sensitized cells and Li-ion battery due to their wide electrochemical window [110]. The high polarity of DESs has led to their potential usage in the purification of biodiesel and biomass valourization. Researchers have explored the potential for molecular aggregation within DESs and DES-based systems. Consequently, self-assembly of numerous surfactants into various aggregated species within DESs has led to their application as media for biocatalysis and drug delivery [111–114]. They are extensively recognized as media for synthesis of various inorganic materials including zeolite-type materials, metal-organic frameworks (MOFs), metal oxides, and specifically shaped nanomaterial. DESs-based systems have been employed as inexpensive and environmentally benign media for CO_2 adsorption and sequestration [115]. The use of DESs in biotransformation and biomedical applications is yet another evolving field of research. DESs have emerged as an excellent choice over traditional organic solvents due to the noteworthy applications of these alternative media in academia and industries. The future offers a significant potential to investigate the

1.4. Aggregation of Cyanine Dyes

The discovery of novel photophysical and photochemical properties of cyanine dyes has attracted the interest of the various scientific community. Cyanine dyes have evoked great curiosity predominantly due to their light absorbing properties arising from the presence of polymethine backbone that involves an extended delocalized π-system [121, 122]. Due to their inherent structural feature, cyanine dyes have an outstandingly high value of extinction coefficients often exceeding 100,000 $Lmol^{-1}cm^{-1}$. The spectrochemical properties such as absorbance/fluorescence transition maxima, color, and photostability are controlled by the length of the polymethine chain of the cyanine dye [123, 124]. For example, cyanines with longer polymethine chain possess higher absorbance/emission wavelengths. Moreover, the geometry of cyanine dyes accounts for their highly sensitive photophysics which allows the effects of variable solvents, temperature, and ionic strength to be characterized. Various cyanine dyes with their trivial names are shown in Figure 5.

One of the peculiar properties of cyanine dyes is their tendency to form self-assemblies stimulated from the strong intermolecular van der Waals attractive forces between the cyanine dye molecules [125]. The phenomenon self-association/aggregation of cyanine dyes often occurs both in aqueous solution as well as on solid surfaces. In 1936, Jelley and Scheibe first witnessed the self-assembled molecular aggregates of cyanine dyes and classified them on the basis of their spatial arrangement as J- and H- type aggregates. Jelly and co-workers further observed that the absorption spectra of cyanine dye significantly reflected the formation of a hypsochromically shifted broad-band and bathochromically shifted narrow-band depicting H- and J-aggregates, respectively (Figure 6) [126–128]. Frank and Teller explained these spectroscopic shifts by molecular exciton coupling theory, i.e., the coupling of transition moments of the constituent dye molecules [129]. According to this theory, the interaction of neighbouring transition dipoles of tightly packed dye molecules leads to splitting of the excited state into two new states referred to as upper and lower exciton levels. A transition to the upper exciton state is allowed for dye molecules oriented in a parallel fashion (plane-to-plane) forming H-aggregates, on the other hand, the transition to the lower state is allowed for a head-to-tail (end-to-end stacking) arrangement resulting in J-aggregates as shown in Figure 6. However, the exciton model is only applicable where interaction between orbitals of constituent dye molecules is negligible [130, 131]. Further, Kasha and co-workers provided an important theory which relates aggregate geometry to the spectroscopic properties of dye molecules [132]. A structure-function relationship in molecular aggregates can be established by a single parameter α (slippage angle), which is the angle form by molecule transition dipole with the axis of aggregates. Large molecular slippage ($\alpha < 32°$) and small slippage ($\alpha > 32°$) result in J-aggregation (bathochromic shift) and H-aggregation (hypsochromic shift) formation, respectively [133].

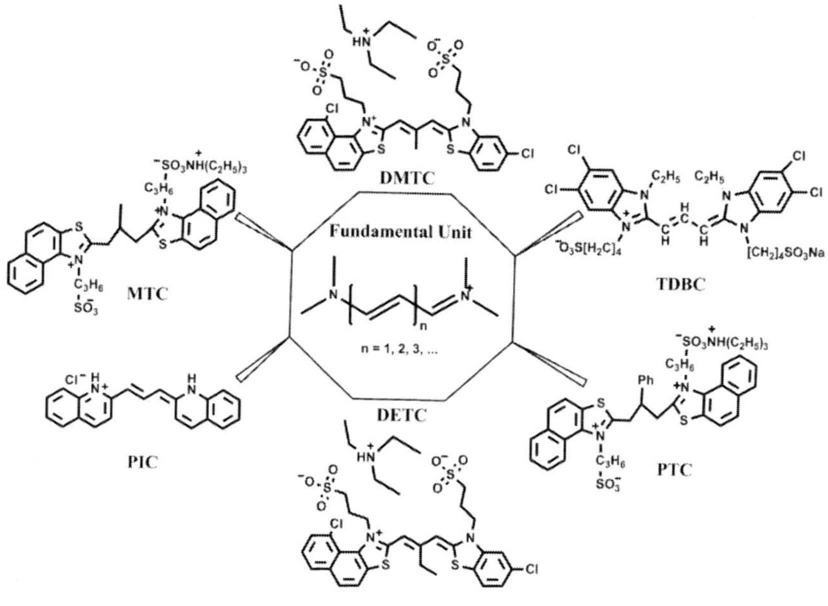

Figure 5. Chemical structures of various cyanine dyes along with their trivial names.

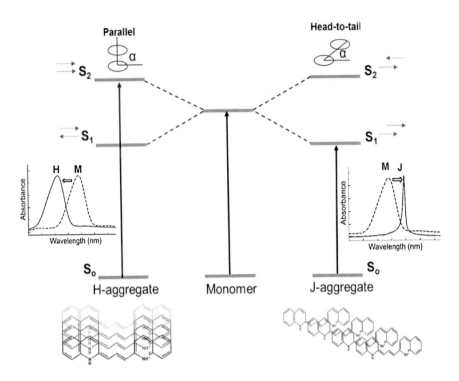

Figure 6. Jablonski diagram of dye aggregation and schematic representation of the changes in absorption spectra on the formation of H- and J-aggregates from cyanine dye.

The inherent chemical structure of the cyanine dye prompts their self-organization into highly ordered aggregates of various structures and morphologies leading to interesting photophysical properties. These unusual properties of cyanine dyes led to their applications in the photographic industry [134]. The advancement of cyanine dye aggregates as potential material for non-linear optics and model compound for photosynthetic pigments created promising applications in many disciplines such as proteomics, optics, photovoltaics, artificial photosynthetic systems, photodynamic therapy of cancer, laser technologies, and markers for biological membrane systems [135, 136]. The importance of cyanine dye has stimulated an enormous amount of scientific work in various areas of science. In the past decades, a large number of publications has been devoted to their vast applications in various fields (Figure 7).

The process of cyanine dye aggregation shows complex dependency on many factors such as solvent polarity, pH, temperature, and structure of the dye [134]. However, the nature of the solvent plays a crucial role in deciding the aggregation behavior of the dye. Furthermore, the external additives like inorganic salts, surfactants, liquid-polymers, electrolytes, and chiral moieties have a substantial effect on the cyanine dye aggregation [137]. Thus, a detailed study of interactions present between solvent and dye monomers is of prime importance to understand the process of aggregation for their further exploration in various fields.

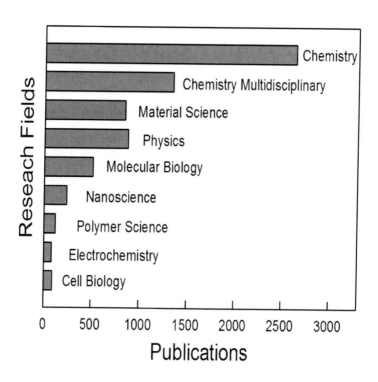

Figure 7. Distribution of the number of publications in cyanine dye (source ISI web of knowledge).

Several reports are available suggesting that water is the most favored media for dye aggregation. This is because of the high dielectric constant value of water results in a reduction of the repulsive forces between the dye molecules, which along with the disruption of the regular hydrogen-bonded structure of water facilitates the process of cyanine dye aggregation [138]. The literature suggests that the concentration of inorganic salt is a major factor in controlling the aggregation behavior of the cyanine dye. Koti and co-workers observed that the addition of these salts initiates the conversion of dye monomer to dimer, higher-order aggregates, H-aggregates, and J-aggregates due to the structural variation in the region around dye molecules [139]. Moreover, a high concentration of inorganic salt facilitates the formation of aggregates and also increases their stability. Xiang JF and co-workers further reported the effect of NaCl on the aggregation behavior of two cyanine dyes- thiacarbocyanine dyes 3,3'-di(3-sulfopropyl)-4,5,4',5-dibenzo-9-phenyl-thiacarbocyanine triethyl ammonium salt (PTC) and 3,3'-di(3-sulfopropyl)-4,5,4',5-dibenzo-9-methyl-thiacarbocyanine triethyl ammonium salt (MTC) within aqueous media [140]. They analyzed the effect of Na^+ ions by employing the ^{1}H-NMR technique and proposed that the ions penetrate within J-aggregates and take over the place of the cationic counterion present in the dye molecule, facilitating the aggregation process of PTC and MTC. Ever since many researchers have investigated the dependence of dye aggregation on the nature of inorganic salt. Chibisov et al. proposed that the inorganic salt stimulates the formation of J-aggregate in water by increasing the dielectric constant value and reducing the static-field repulsion among the monomers of dye [141]. Moreover, they found that the mode of assembly in dye molecules can be controlled and regulated by the identity of metal ion present in the salt. The authors also observed that the rate of J-aggregation follows the order: trivalent > divalent > monovalent metal ion. Besides inorganic salt, chelating compounds have been observed to show some interesting outcomes. An addition of chelating molecule to the salt-added system results in the formation of H-aggregate due to segregation of the metal ion from the dye molecules whereas, without chelating compound, inorganic salt favors the formation of J-aggregates [142].

Thus, it can be stated that the aggregation behavior of cyanine dye is largely sensitive to the nature of the environment surroundings [143].

Deeper studies on the interplay between dye molecules in different solvents are required in order to gain insight into the geometry and photophysical properties of aggregates, which will further help open new perspectives for creating an efficient application in all fields [144–147]. As mentioned earlier ILs and DESs have shown the remarkable application in various aggregation processes due to their unique inherent structure and significant properties. Therefore, studying the aggregation behavior of cyanine dyes in these novel and alternate media is of great interest in approaching the goal of green chemistry [148]. With an aim to inspire further advances, this chapter is focused on the characteristics of cyanine dye aggregation within IL- and DES-based systems.

2. AGGREGATION WITHIN IONIC LIQUID BASED SYSTEMS

ILs have been established as supporting media for the process of aggregation of various molecules [149]. The anomalous behavior of ILs stems from the fact that they are constituting entirely of ions. The embedded electronic structure of ILs and their unique physicochemical properties have evoked an area of research in the dye aggregation process. Various spectroscopic techniques like uv-vis absorption spectroscopy, fluorescence emission spectroscopy, and resonance light scattering (RLS) have been employed by researchers to investigate the cyanine dye aggregation within ILs. In the present chapter, we provide detailed information about the dye aggregation process within ILs.

2.1. Ionic Liquids as Solubilizing Media

Many researchers have noticed the unusual aspects of ILs as solubilizing media for dye aggregation process, though the reports are few and scarce. In one of the investigation, Kumar and co-workers demonstrated the aggregation behavior of two anionic cyanine dyes; 5,5′-dichloro-3,3′-di(3-sulfopropyl)-9-methyl-thiacarbocyanine triethylammonium salt (DMTC) and 5,5′,6,6′-tetrachloro-1,1′-diethyl-3,3′-di(4-sulfobutyl) benzimidazolocarbocyanine sodium salt (TDBC) [150]. They studied the self-aggregation of these cyanine dyes in various ILs; 1-butyl-3 methylimidazolium tetrafluoroborate ([bmim][BF$_4$]), 1-butyl-3 methylimidazolium hexafluorophosphate ([bmim][PF$_6$]) and 1-hexyl-3-methylimidazolium tetrafluoroborate ([hmim][BF$_4$]). The group reported that cyanine dyes TDBC and DMTC were prompted to form fluorescent H-aggregates in the presence of 2 wt% of 1 M aqueous NaOH in ILs having [bmim] and [hmim]-based cations and BF$_4^-$ as an anion. On the other hand, they found unusual aggregation behavior of TDBC within [bmim][PF$_6$], wherein J-aggregates are reported to form upon addition of 2 wt% of 1 M aqueous NaOH [150]. These different outcomes are attributed to the anion dependent hydrolytic properties of ILs for example, BF$_4^-$ as anion results in hydrolytically unstable nature of IL whereas PF$_6^-$ anion constitutes hydrolytically stable ILs. Thus, the identity of ILs governs the aggregation behavior of TDBC and DMTC.

H-aggregates of TDBC can be characterized by a hypsochromically-shifted absorbance band at 440 nm besides M-band (monomer) present at 520 nm [150]. Interestingly, the authors observed that the addition of a small amount of basic water can initiate the formation of H-aggregates in the [bmim][BF$_4$]. However, the efficiency and kinetics of H-aggregation of TDBC are found to be maximum in 4 wt % of 1 M aqueous NaOH in [bmim][BF$_4$] (Figure 8). More than 4 wt% of NaOH results in lower aggregation efficiency as well as decreased kinetics of the H-aggregation. Whereas, the addition of 2 wt% of 1 M aqueous NaOH to TDBC solution in [bmim][PF$_6$] has shown bathochromically-shifted (J-band) band at 586 nm, attributed to the presence of J-aggregates, along with the M-band at 520 nm in uv-vis absorbance spectra as shown in Figure 8.

Excitation of TDBC in [bmim][BF$_4$] with 4 wt% NaOH (1 M) results in the appearance of a new band at 465 nm (H-band) besides the monomer band centered at 540 nm, thus, fluorescent H-aggregates of TDBC can be characterized by hypsochromaclly-shifted band by fluorescence study (Figure 9).

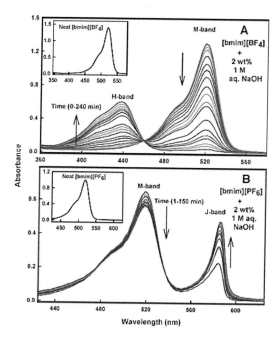

Figure 8. Change in absorbance spectra of TDBC in aqueous NaOH added-[bmim][BF$_4$](Panel A) and aqueous NaOH added-[bmim][PF$_6$] (Panel B) with time at ambient conditions. Insets show TDBC in neat ILs -[bmim][BF$_4$] and [bmim][PF$_6$]. [Reprinted (adapted) with permission from Kumar and co-workers (2011) Chem. Commun. 4730−4732. Copyright (2011), Royal Society of Chemistry] [150].

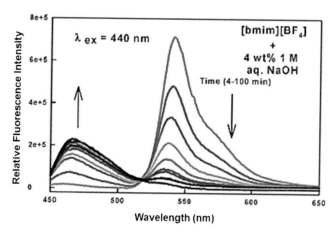

Figure 9. Fluorescence spectra of TDBC in aqueous NaOH added [bmim][BF$_4$] with time. [Reprinted (adapted) with permission from Kumar and co-workers (2011) Chem. Commun. 4730−4732. Copyright (2011), Royal Society of Chemistry] [150].

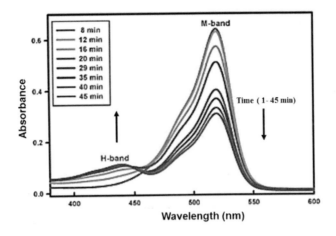

Figure 10. Change in absorption spectra of TDBC in aqueous NaOH added-[bmim][BF$_4$] with time at 100 °C. [Reprinted (adapted) with permission from Kumar and co-workers (2011) Chem. Commun. 4730–4732. Copyright (2011), Royal Society of Chemistry] [150].

Contrary to the previous studies where H-aggregates are reported to be non-fluorescent, this group reported the fluorescent nature of H-aggregates. Similar results are obtained for TDBC aggregation behavior in IL composed of [hmim]$^+$ and [BF$_4$]$^-$. J-aggregates of TDBC dye were also observed when it was dissolved in other ILs; [emim][Tf$_2$N], [bmim][OTf], [bmim][Tf$_2$N], and [bmpyrr][Tf$_2$N] (1-butyl-1-methylpyrrolidinium bis(trifluoromethanesulfonyl)imide) with 2 wt% 1 M aqueous NaOH at the ambient conditions [150]. Fluorescent features of J-aggregates were confirmed by using fluorescence emission spectra of TDBC in [bmim][PF$_6$] with 2 wt% of 1 M aqueous NaOH. Furthermore, J-aggregates were characterized by a smaller Stoke's shift (> 3 nm) relative to H-aggregates (25 nm) in fluorescence emission spectra of TDBC. The group investigated the aggregation behavior of DMTC in [bmim][PF$_6$] and [hmim][BF$_4$] with 2 wt% of 1 M aqueous NaOH and found the aggregation behavior to be similar to that of TDBC. To understand the effect of temperature on aggregation behavior of cyanine dyes in ILs, this analysis were performed at a higher temperature (100 °C). It is interesting to note that the high temperature favors the formation of H-aggregates in basic water-added ILs irrespective of the identity of the IL (Figure 10).

Moreover, no J-band was observed at higher temperature implying the disruption of J-aggregates with an increase in temperature.

2.2. Ionic Liquids as Additives

Kumar and co-workers also studied the phenomena of conversion of J-aggregates of cyanine dyes into monomers in the aqueous system by using hydrophilic ILs as additives [151]. They investigated the concentration-dependent dual character of ILs, where at the lower concentrations, they act as inorganic salts and co-solvents at higher concentration.

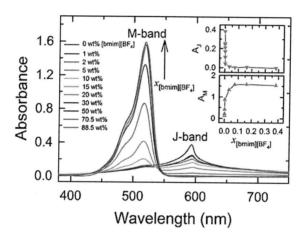

Figure 11. Absorbance spectra of TDBC in the presence of the varying amount of [bmim][BF$_4$] in high pH PB solution. Inset represents a change in J- and M-band absorbance with the amount of [bmim][BF$_4$]. [Reprinted (adapted) with permission from Kumar and co-workers (2011) Langmuir 27:12884−12890. Copyright (2011), American Chemical Society] [151].

The authors reported that TDBC gets decolourized in the presence of neutral/low-pH water, as a result, no aggregation was observed, whereas, at higher pH, ~11-12 TDBC forms J-aggregates. To investigate the effect of [bmim][BF$_4$] as additive, amount of [bmim][BF$_4$] was varied from 5 to 10 wt% in high pH water. It was observed that the addition of the [bmim][BF$_4$] initiates the conversion of J-aggregates into monomeric species of TDBC [151]. This effect became more significant as the amount of [bmim][BF$_4$] increased from 10 to 50 wt% (Figure 11), where all the J-aggregates converted into monomers. In contrast, the addition of common salts, such as NaBH$_4$, NaPF$_6$, and NaCl failed to initiate this conversion of J-aggregates into monomers. The disintegration of J-aggregates into monomers within [bmim][BF$_4$]-aqueous system is a result of the increment in the repulsive forces between the TDBC dye monomers as well as reduction of the dielectric constant of the media. The authors also proposed that on the addition of [bmim][BF$_4$] to basic buffer solution of TDBC, [bmim]$^+$ interacts with TDBC J-aggregates via π-π interaction leading to their disruption [151].

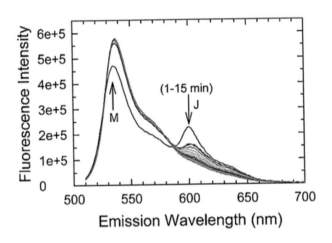

Figure 12. Change in fluorescence emission spectra of TDBC with time within pH = 12 PB solution in the presence of [bmim][BF$_4$]. [Reprinted (adapted) with permission from Kumar and co-workers (2011) Langmuir 27:12884−12890. Copyright (2011), American Chemical Society] [151].

To further see the effect of time and pH on the conversion phenomena of TDBC, the group carried out the time-dependent absorbance studies of TDBC in 20 wt% [bmim][BF$_4$] PB solution at pH 8.5 and 12. They found maximum conversion of J-aggregates into monomers (Monomer absorbance-A$_M$ higher than J-aggregate absorbance-A$_J$) in case of pH = 12 as compared to pH = 8.5 (Figure 12).

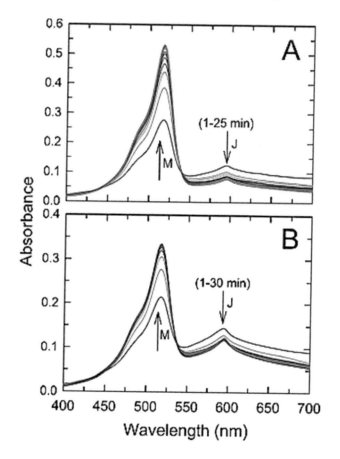

Figure 13. Change in absorption spectra of TDBC with time within pH = 12 PB solution in the presence of [bmim][OTf] (Panel A) and [bmpyrr][OTf] (Panel B). [Reprinted (adapted) with permission from Kumar and co-workers (2011) Langmuir 27:12884−12890. Copyright (2011), American Chemical Society] [151].

The conversion of TDBC J-aggregates into monomer species upon addition of [bmim][BF$_4$] was validated by using the fluorescence technique. Time-dependent fluorescence emission studies of TDBC in basic solution in the presence of 20 wt% of [bmim][BF$_4$] was carried out (Figure 13). It was observed that on increasing time, the emission intensity of J-aggregate decreased gradually with a parallel growth in the monomer emission band indicating the changeover of J-aggregates into monomer units over a span of time. In expansion, the authors further investigated the structural effect of IL on the interconversion of TDBC J-aggregates into monomers using two different ILs ([bmim][OTf] and [bmpyrr][OTf]). Interestingly, these two ILs displayed behavior similar to [bmim][BF$_4$]. Some water immiscible ILs were also used to investigate the TDBC aggregation within them by the authors. The IL [bmim][PF$_6$] which has lower water miscibility (2.0 wt% solubility) showed similar results, whereas in case of [bmim][PF$_6$] no recovery of monomer was observed.

From the outcomes of investigations, it is evident that IL can have a substantial effect on the aggregation behavior of cyanine dyes. IL-based media supports the aggregation of cyanine dyes. Furthermore, the recovery of TDBC monomers from their J-aggregates upon the addition of ILs to the aqueous system is aptly highlighted. The dual nature of ILs as inorganic salts and co-solvents, at lower and higher concentrations, respectively is demonstrated.

3. AGGREGATION WITHIN DEEP EUTECTIC SOLVENT BASED SYSTEMS

DESs possess significant dipolarities arising from their unique structure that features extensive H-bonding ability. Interestingly, this inherent structural aspect has been reported to trigger self-aggregation of various molecules and sustain supramolecular assemblies within DESs. Similar to ILs, DESs can also be envisioned to influence self-aggregation of the cyanine dyes in an unusual manner which could help create interesting applications for this novel non-aqueous media. The self-association of cyanine dyes in aqueous solution is an extensively studied phenomenon by researchers. However, not much has been reported in the literature about their aggregation within DESs-based media and therefore it is a prospective area for extensive research. Pal and co-workers carried out the first and only investigation of cyanine dye within DESs, wherein, a detailed photophysical study on aggregation behavior of TDBC and DMTC dye within DES-based media is reported [152].

3.1. Neat and Aqueous NaOH-Added DESs

Pal and co-workers used ChCl-based DESs- Glyceline and Reline, which are obtained by mixing ChCl with HBDs, glycerol and urea, respectively, in 1 : 2 molar ratio. They observed a single monomeric band at 527 nm in uv-vis spectra of TDBC in neat Glyceline and Reline, unlike in water where most of its uv-vis radiation absorbing efficiency is lost due to protonation of the dye [152]. However, they observed that the addition of aqueous NaOH to TDBC solution in DES triggers the formation of J-aggregates (J-band at 594 nm) at the expense of the monomer band (M-band) (Figure 14). The authors performed several experiments to optimize the concentration of aqueous NaOH in DES for dye aggregation and found efficient aggregate formation at 1.7 M NaOH in 30 wt% water. The rate of J-aggregation is known to depend on the HBD acidity and dipolarity/polarizability of media. The higher value of J- to M-band absorbance maxima ratio (A_J/A_M) for aqueous NaOH-added TDBC solution in Glyceline (2.80) than Reline (2.06) is attributed to the higher acidity of glycerol [152]. The observed aggregation phenomenon is similar for hydrolytically-stable ILs, e.g., [bmim][PF$_6$] [150]. On the contrary, hydrolytically-unstable ILs (e.g., [bmim][BF$_4$]) show a completely different aggregation behavior [150]. Further, the emission spectra clearly established the fluorescent nature of the J-aggregates formed in Glyceline and Reline in the presence of aqueous NaOH (Figure 14). The formation of aggregates resulted in decreased fluorescence lifetime (subnanosecond) compared to that for neat DESs (1.2 – 1.4 ns).

Similarly, for DMTC in neat Glyceline/Reline, the only monomer band (at 550 nm) was observed in uv-vis absorbance spectra of the carbocyanine dye. However, in the presence of aqueous NaOH, DMTC forms H- as well as J-aggregate which can be characterized by the appearance of absorption bands at 400 and 610 nm, respectively (Figure 15).

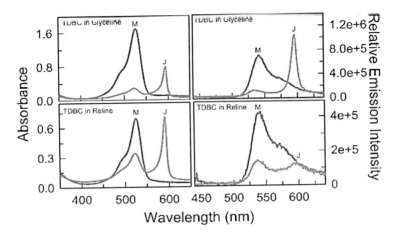

Figure 14. Absorbance and fluorescence emission spectra of TDBC dissolved in neat Glyceline/Reline (black) and upon addition of aqueous NaOH to Glyceline/Reline (grey). [Reprinted (adapted) with permission from Pal and co-workers (2017) Langmuir 33:9781−9792. Copyright (2017), American Chemical Society] [152].

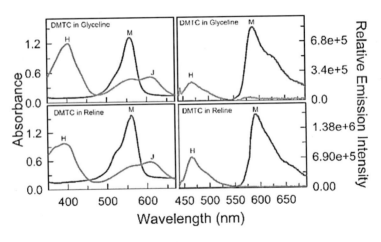

Figure 15. Absorbance and fluorescence emission spectra of DMTC dissolved in neat Glyceline/Reline (black) and upon addition of aqueous NaOH to Glyceline/Reline (grey). [Reprinted (adapted) with permission from Pal and co-workers. (2017) Langmuir 33:9781−9792. Copyright (2017), American Chemical Society] [152].

It was observed that the J-band appeared instantly after the addition of aqueous NaOH to Glyceline/Reline. However, the band corresponding to H-aggregates was seen to grow at the expense of M- and J-band showing the conversion of monomers and J-aggregates to H-aggregates over a period of 90 min [152]. It is interesting to note that the H-aggregates formed in aqueous NaOH-added DMTC solution in Glyceline/Reline exhibit fluorescence whereas no fluorescence is observed for H-aggregates of DMTC in basic water. Unlike DMTC in aqueous NaOH-added Glyceline/Reline, J-aggregates are not formed at all in basic water-added DMTC. A stark difference in the aggregation behavior of TDBC versus DMTC can be explained on the basis of their structural dissimilarities [152]. As observed in the case of TDBC J-aggregates, the H-aggregates formed by DMTC in aqueous-base added Glyceline/Reline also have shorter lifetimes.

3.2. Surfactant-Added DESs

The photophysical behavior of cyanines is known to be affected by the presence of surfactants (as monomers or micelle). Therefore, the study of cyanine aggregation in surfactant-added media has become a subject of great interest. In an attempt to explore the effect of surfactants on dye aggregation, the authors assessed the aggregation behavior of zwitterionic TDBC and DMTC dyes in presence of cationic surfactant cetyltrimethylammonium bromide (CTAB) and anionic surfactant sodium dodecyl sulfate (SDS) [152]. However, data acquisition was not possible for CTAB in Reline due to its solubility restriction.

For TDBC, it was observed that on the addition of very low CTAB concentration ~ 0.1 mM (below Critical Micelle Concentration, CMC) to Glyceline, the surfactant monomers aids the J-aggregate formation. However, no J-band is observed at 1 mM CTAB concentration (below CMC). Nonetheless, addition of aqueous NaOH ([NaOH] = 1.7 M, water = 30 wt%) to 1 mM CTAB-added DES system results in spontaneous J-aggregation of TDBC. Interestingly, increasing the concentration of CTAB above CMC in aqueous NaOH-added Glyceline results in the disruption of J-aggregates and formation of unprecedented H-aggregates of the dye [152]. However, the H-aggregation within micellar CTAB in the presence of NaOH is not as efficient in Glyceline as that in water. The authors attributed this outcome to the shape/size of micellar CTAB which may not be very favorable to H-aggregate formation/accommodation in aqueous NaOH-added Glyceline. It is to be noted that the aggregates of TDBC formed in aqueous base-added micellar CTAB solution of Glyceline are fluorescent in nature. The results were very similar for SDS, as no aggregation was observed at surfactant concentration below CMC (1 mM SDS). However, the J-band readily evolved at the cost of the M-band in both Glyceline and Reline on the addition of aqueous NaOH. Due to partial solidification, authors could not carry out data acquisition for SDS in the micellar form in DESs [152]. It was highlighted that J-aggregation efficiency in the presence of SDS is more pronounced in aqueous NaOH-added Glyceline as compared to aqueous NaOH-added Reline (Figure 16). This is similar to what was observed for NaOH-added DESs. The J- and H-aggregates of TDBC formed by the addition of aqueous NaOH to surfactant-added DES are fluorescent in nature.

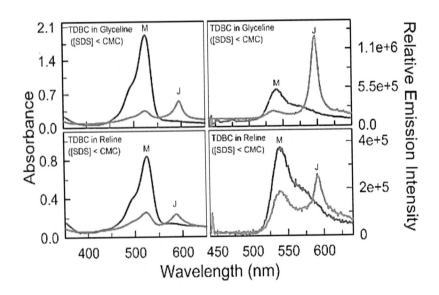

Figure 16. Absorbance and fluorescence emission spectra of TDBC dissolved in SDS-added Glyceline/Reline (black) and upon addition of aqueous NaOH to SDS-added Glyceline/Reline (grey). [Reprinted (adapted) with permission from Pal and co-workers (2017) Langmuir 33:9781−9792. Copyright (2017), American Chemical Society] [152].

For DMTC, unlike water, no aggregation was prompted in CTAB or SDS-added neat DESs. However, the addition of aqueous NaOH to CTAB-added Glyceline facilitated nearly the complete conversion of DMTC monomers into H-aggregates. DMTC exclusively forms H-aggregates irrespective of the CTAB concentration in aqueous NaOH-added Glyceline, unlike base-added Glyceline which contains appreciable monomer and J-aggregates along with the H-aggregates. The weakly fluorescent H-aggregates of DMTC are formed within aqueous base-added CTAB solution of Glyceline whereas the H-aggregates are not fluorescent for CTAB solution in water. Moreover, the authors noted quite interesting results for DMTC in aqueous NaOH-added SDS solution of DES [152]. The aggregation pattern of this carbocyanine dye in the presence of monomeric SDS is controlled by the identity of the DES. In the corresponding Glyceline solution, the M-band diminished and a weak shoulder representing J-aggregates at 610 nm appeared along with a prominent peak for H-aggregate. Whereas for Reline, both the J-band and the H-band are clearly observed along with the M-band as a shoulder. The H-aggregates so formed are found to be highly fluorescent in nature.

From the overall study, it was clear that DES-based systems can potentially support the self-aggregation of the carbocyanine dyes TDBC and DMTC into J- or H-aggregate depending on nature and concentration of additives present.

CONCLUSION

ILs and DESs are emerging as the versatile green solvents. The unique architecture and properties of these solvents provide them an advantage of having promising solvent qualities. This chapter provides a detailed overview of cyanine dye aggregation in IL and DES based systems. From the overall study, it is clear that though neat ILs/DESs do not show any formation of aggregates, the additives trigger the self-aggregation of the carbocyanine dyes TDBC and DMTC into J- or H-aggregate. Exceptionally, fluorescent H-aggregates of TDBC and DMTC are formed within IL/DES based media which are quite uncommon to be observed in any other media. This is attributed to the unique inherent architecture of ILs and DESs. Also, the dye aggregation efficiency within IL/DES based media is quite different from that observed in corresponding aqueous basic media. The dependence of cyanine dye aggregation behavior on the different constituents of ILs/DESs is highly emphasized. Hence, the outcomes amply support the growing potential of these green solvents as alternate media for molecular aggregation processes which opens up new avenues for cyanine dye applications in various fields of science. However, only a narrow range of ILs and DESs have been investigated so far. Therefore, there is a need to explore different classes of DESs and ILs in order to get a better insight of cyanine dye aggregation within these media.

ACKNOWLEDGMENTS

The works of the authors presented in this article were generously supported by the Science and Engineering Board (SERB) and Council of Scientific and Industrial Research (CSIR) of Government of India.

REFERENCES

[1] Reichardt, C. 2004. "Pyridinium N-phenolate Betaine Dyes as Empirical Indicators of Solvent Polarity: Some New Findings." *Pure and Applied Chemistry* 76: 1903–1919. doi:10.1351/pac200476101903.

[2] Reichardt, C. 1994. "Solvatochromic Dyes as Solvent Polarity Indicators." *Chemical Reviews* 94: 2319–2358. doi:10.1021/cr00032a005.

[3] Reichardt, C. 2003. *Solvents and Solvent Effects in Organic Chemistry*. Weinheim: Wiley-VCH.

[4] Buncel, E., R. A. Stairs, and H. Wilson. 2003. *The Role of the Solvent in Chemical Reactions.* Oxford: Oxford University Press.

[5] Wypych, G. 2001. *Handbook of Solvents.* Toronto: ChemTec Publishing.

[6] Zhang, X., M. M. Cunningham, and R. A. Walker. 2003. "Solvent Polarity at Polar Solid Surfaces: The Role of Solvent Structure." *The Journal of Physical Chemistry B* 107: 3183–3195. doi:10.1021/jp021071i.

[7] Yang, E. S., D. M. Cunnold, M. J. Newchurch, and R. Salawitch. 2005. "Change in Ozone Trends at Southern High Latitudes." *Geophysical Research Letters.* 32: L12812. doi:10.1029/2004GL 022296.

[8] McCulloch, A. 2003. "Fluorocarbons in the Global Environment: A Review of the Important Interactions with Atmospheric Chemistry and Physics." *Journal of Fluorine Chemistry* 123: 21–29. doi:10. 1016/S0022-1139(03)00105-2.

[9] Morgenstern, O., P. Braesicke, M. M. Hurwitz, F. M. O'Connor, A. C. Bushell, C. E. Johnson, and J. A. Pyle. 2008. "The World Avoided by the Montreal Protocol." *Geophysical Research Letters* 35: L16811. doi:10.1029/2008GL034590.

[10] Kumar, S. S., and D. Subhashini. 2015. "Health Hazards of Organic Solvents." *Research and Reviews: Journal of Chemistry* 4: 90–95. http:// www.rroij.com/ open- access/ health- hazards- of- organic-solvents.pdf.

[11] Omer, A. M. 2008. "Energy, Environment and Sustainable Development." *Renewable and Sustainable Energy Reviews* 12: 2265–2300. doi:10.1016/j.rser.2007.05.001.

[12] Goharshadi, E. K., Y. Ding, and P. Nancarrow. 2008. "Green Synthesis of ZnO Nanoparticles in a Room-Temperature Ionic Liquid 1-ethyl-3-methylimidazolium bis(trifluoromethylsulfonyl) imide." *Journal of Physics and Chemistry of Solids* 69: 2057–2060. doi:10.1016/j.jpcs.2008.03.002.

[13] Perosa, A., and F. Zecchini. 2007. *Methods and Reagents for Green Chemistry: An Introduction.* Translated by Pietro Tundo. New York: *John Wiley & Sons.*

[14] Kunle, O. O., J. M. Fortunak, and R. D. Rogers. 2008. "Workshop in Green Chemistry Production of Essential Medicines in Developing Countries." *Green Chemistry* 10: 823–824. doi:10.1039/B806001K.

[15] Armenta, S., S. Garrigues, and M. de la Guardia. 2008. "Green Analytical Chemistry." *Trends in Analytical Chemistry* 27: 497–511. doi:10.1016/j.trac.2008.05.003.

[16] Clarke, C. J., W. C. Tu, O. Levers, A. Bröhl, and J. P. Hallett. 2018. "Green and Sustainable Solvents in Chemical Processes." *Chemical Reviews* 118: 747–800. doi:10.1021/acs.chemrev.7b00571.

[17] Leitner, W., and M. Poliakoff. 2008. "Supercritical Fluids in Green Chemistry." *Green Chem*istry 10: 730. doi:10.1039/B809498P.

[18] Abaee, S. M., M. M. Mojtahedi, H. Abbasi, and E. R. Fatemi. 2008. "Additive-Free Thiolysis of Epoxides in Water: A Green and Efficient Regioselective Pathway to β-Hydroxy Sulfides." *Synthetic Communications* 38: 282–289. doi:10.1080/00397910701749963.

[19] Gabriel, S., and J. Weiner. 1888. "About Some Derivatives of Propylamine." *European Journal of Inorganic Chemistry Communication* 21: 2669–2679. doi:10.1002/cber.18880210288.

[20] Wilkes, J. S., J. A. Levisky, R. A. Wilson, and C. L. Hussey. 1982. "Dialkylimidazolium Chloroaluminate Melts: A New Class of Room-Temperature Ionic Liquids for Electrochemistry, Spectroscopy and Synthesis." *Inorganic Chemistry* 21: 1263–1264. doi:10.1021/ic00133a078.

[21] Wilkes, J. S., and M. J. Zaworotko. "Air and Water Stable 1-Ethyl-3-Methylimidazolium Based Ionic Liquids." *Journal of the Chemical Society, Chemical Communications* 0: 965–967. doi:10.1039/c39920000965.

[22] Freemantle, M. 1998. "Designer Solvents." *Chemical & Engineering News* 76: 32–37. doi:10.1021/cen-v076n013.p032.

[23] Newington, I., J. M. Perez-Arlandis, and T. Welton. 2007. "Ionic Liquids as Designer Solvents for Nucleophilic Aromatic

Substitutions." *Organic Letters* 9: 5247–5250. doi:10.1021/ol702435f.

[24] Behera, K., P. Dahiya, and S. Pandey. 2007. "Effect of Added Ionic Liquid on Aqueous Triton X-100 Micelles." *Journal of Colloid and Interface Science* 307: 235–245. doi:10.1016/j.jcis.2006.11.009.

[25] Wasserscheid, P., and T. Welton. 2002. *Ionic Liquids in Synthesis.* Germany: Wiley-VCH Verlag, Stuttgart.

[26] Ghandi. K. 2014. "A Review of Ionic Liquids, Their Limits and Applications." *Green and Sustainable Chemistry* 4: 44–53. doi:10.4236/gsc.2014.41008.

[27] Weingarth, D., I. Czekaj, Z. Fei, A. Foelske-Schmitz, P. J. Dyson, A. Wokaun, and R. Kötz. 2012. "Electrochemical Stability of Imidazolium Based Ionic Liquids Containing Cyano Groups in the Anion: A Cyclic Voltammetry, XPS and DFT Study." *Journal of the Electrochemical Society* 159: 611–615. doi: 10.1149/ 2.001207jes.

[28] Wang, C., H. Luo, X. Luo, H. Li, and S. Dai. 2010. "Equimolar CO_2 Capture by Imidazolium-Based Ionic Liquids and Superbase Systems," *Green Chemistry* 12: 2019–2023. doi:10.1039/c0gc00070a.

[29] Safaei-Ghomi, J., M. Emaeili, and R. H. Abdol. 2010. "Mild and Efficient Method for Oxidation of Alcohols in Ionic Liquid Media." *Digest Journal of Nanomaterials and Biostructures* 5: 865–871. http://www.chalcogen.ro/865_Ghomi1.pdf.

[30] Yu, J. I., H. Y. Ju, K. H. Kim, and D. W. Park. 2010. "Cycloaddition of Carbon Dioxide to Butyl Glycidyl Ether Using Imidazolium Salt Ionic Liquid as a Catalyst." *Korean Journal of Chemical Engineering* 27: 446–451. doi:10.1007/s11814-010-0074-1.

[31] Sarkar, A., S. R. Roy, N. Parikh, and A. K. Chakraborti. 2011. "Nonsolvent Application of Ionic Liquids: Organo-Catalysis by 1-alkyl-3-methylimidazolium Cation Based Room-Temperature Ionic Liquids for Chemoselective N-tert-Butyloxycarbonylation of Amines and the Influence of the C-2 Hydrogen on Catalytic Efficiency." *The Journal of Organic Chemistry* 76: 7132–7140. doi:10.1021/jo201102q.

[32] Canal, J. P., T. Ramnial, D. A. Dickie, and J. A. C. Clyburne. 2006. "From the Reactivity of N-Heterocyclic Carbenes to New Chemistry in Ionic Liquids." *Chemical Communications* 2006: 1809–1818. doi:10.1039/b512462j.

[33] Aggarwal, V. K., I. Emme, and A. Mereu. 2002. "Unexpected Side Reactions of Imidazolium-Based Ionic Liquids in the Base-Catalysed Baylis–Hillman Reaction." *Chemical Communications* 2002: 1612–1613. Doi:10.1039/b203079a.

[34] Khupse, N. D., and A. Kumar. 2011. "The Cosolvent-Directed Diels-Alder Reaction in Ionic Liquids." *The Journal of Physical Chemistry A* 115: 10211–10217. doi:10.1021/jp205181e.

[35] Snelders, D. J. M., and P. J. Dyson. 2011. "Efficient Synthesis of β-Chlorovinylketones from Acetylene in Chloroaluminate Ionic Liquids." *Organic Letters* 13: 4048–4051. doi:10.1021/ol201182t.

[36] Ford, L., F. Atefi, R. D. Singer, and P. J. Scammells. 2011. "Grignard Reactions in Pyridinium and Phosphonium Ionic Liquids." *European Journal of Organic Chemistry* 2011: 942–950. doi: 10.1002/ejoc.201001468.

[37] Wu, X. Y. 2011. "Facile and Green Synthesis of 1,4-Dihydropyridine Derivatives in n-Butyl Pyridinium Tetrafluoroborate." *Synthetic Communications* 42: 454–459. doi:10.1080/00397911.2010.525773.

[38] Hajipour, A. R., and M. Seddighi. 2011. "Pyridinium-Based Brønsted Acidic Ionic Liquid as a Highly Efficient Catalyst for One-Pot Synthesis of Dihydropyrimidinones." *Synthetic Communications* 42: 227–235. doi:10.1080/00397911.2010. 523488.

[39] Pajuste, K., A. Plotniece, K. Kore, L. Intenberga, B. Cekavicus, D. Kaldre, G. Duburs, and A. Sobolev. 2011. "Use of Pyridinium Ionic Liquids as Catalysts for the Synthesis of 3,5-Bis(dodecyloxycarbonyl)-1,4-dihydropyridine Derivative." *Central European Journal of Chemistry* 9: 143–148. doi:10.2478/s11532-010-0132-x.

[40] Tsunashima, K., A. Kawabata, M. Matsumiya, S. Kodama, R. Enomoto, M. Sugiya, and Y. Kunugi. 2011. "Low Viscous and

Highly Conductive Phosphonium Ionic Liquids Based on Bis(fluorosulfonyl)amide Anion as Potential Electrolytes." *Electrochemistry Communications* 13: 178–181. doi:10.1016/j.elecom.2010.12.007.

[41] Luska, K. L., K. Z. Demmans, S. A. Stratton, and A. Moores. 2012. "Rhodium Complexes Stabilized by Phosphine Functionalized Phosphonium Ionic Liquids Used as Higher Alkene Hydroformylation Catalysts: Influence of the Phosphonium Headgroup on Catalytic Activity." *Dalton Transactions* 41: 13533–13540. doi:10.1039/c2dt31797d.

[42] Fan, A., G. K. Chuah, and S. Jaenicke. 2012. "Phosphonium Ionic Liquids as Highly Thermal Stable and Efficient Phase Transfer Catalysts for Solid-Liquid Halex Reactions." *Catalysis Today* 198: 300–304. doi:10.1016/j.cattod.2012.02.063.

[43] Harper, N. D., N. D. Nizio, A. D. Hendsbee, J. D. Masuda, K. N. Robertson, L. J. Murphy, M. B. Johonson, C. C. Pye, and J. A. C. Clyburne. 2011. "Survey of Carbon Dioxide Capture in Phosphonium-Based Ionic Liquids and End Capped Polyethylene Glycol Using DETA (DETA = Diethylenetriamine) as a Model Absorbent." *Industrial & Engineering Chemistry Research* 50: 2822–2830. doi:10.1021/ie101734h.

[44] Ghandi, K. 2012. *Process for the Production of Polystyrene and Novel Polymers in Phosphonium Ionic Liquids.* US Patent: 20,120,049,101.

[45] Cheng, S., M. Zhang, T. Wu, S. T. Hemp, B. D. Mather, R. B. Moore, and T. E. Long. 2012. "Ionic Aggregation in Random Copolymers Containing Phosphonium Ionic Liquid Monomers." *Journal of Polymer Science Part A: Polymer Chemistry* 50: 166–173. doi:10.1002/pola.25022.

[46] Lauzon, J. M., D. J. Arseneau, J. C. Brodovitch, J. A. C. Clyburne, P. Cormier, B. McCollum, and K. Ghandi. 2008. "Generation and Detection of the Cyclohexadienyl Radical in Phosphonium Ionic Liquids." *Physical Chemistry Chemical Physics* 10: 5957–5962. doi:10.1039/b804800b.

[47] Taylor, B., P. J. Cormier, J. M. Lauzon, and K. Ghandi. 2008. "Investigating the Solvent and Temperature Effects on the Cyclohexadienyl Radical in an Ionic Liquid." *Physica B: Condensed Matter* 404: 936–939. doi:10.1016/j.physb.2008.11.224.

[48] Bradaric, C. J., A. Downard, C. Kennedy, A. J. Robertson, and Y. Zhou. 2003. "Industrial Preparation of Phosphonium Ionic Liquids." *Green Chemistry* 5: 143–152. doi:10.1039/b209734f.

[49] Dake, S. A., R. S. Kulkarni, V. N. Kadam, S. S. Modani, J. J. Bhale, S. B. Tathe, and R. P. Pawar. 2009. "Phosphonium Ionic Liquid: A Novel Catalyst for Benzyl Halide Oxidation." *Synthetic Communications* 39: 3898–3904. doi:10.1080/0039791090284 0835.

[50] Tseng, M. C., H. C. Kan, and Y. H. Chu. 2007. "Reactivity of Trihexyl(tetradecyl)phosphonium Chloride, a Room-Temperature Phosphonium Ionic Liquid." *Tetrahedron Letters* 48: 9085–9089. doi:10.1016/j.tetlet.2007.10.131.

[51] Cao, H., and H. Alper. 2010. "Palladium-Catalyzed Double Carbonylation Reactions of o-Dihaloarenes with Amines in Phosphonium Salt Ionic Liquids." *Organic Letters* 12: 4126–4129. doi:10.1021/ol101714p.

[52] Fannin Jr., A. A., D. A. Floreani, L. A. King, J. A. Landers, B. J. Piersma, D. J. Stech, R. L. Vaughn, J. S. Wilkes, and J. L. Williams. 1984. "Properties of 1,3-Dialkylimldazollum Chloride-Aluminum Chloride Ionic Liquids. 2. Phase Transitions, Densities, Electrical Conductivities, and Viscosities." *The Journal of Physical Chemistry* 88: 2614–2621. doi:10.1021/j150656a038.

[53] Hu, Y.-F., and C.-M. Xu. 2006. "Effect of the Structures of Ionic Liquids on Their Physical-Chemical Properties and the Phase Behavior of Mixtures Involving Ionic Liquids." *Chemical Reviews*. doi:10.1021/cr0502044.

[54] Bônhote, P., A. P. Dias, N. Papageorgiou, K. Kalyansundaram, and M. Grätzel. 1996. "Hydrophobic, Highly Conductive Ambient-Temperature Molten Salts." *Inorganic Chemistry* 35: 1168–1178. doi:10.1021/ic951325x.

[55] Mutch, M. L., and J. S. Wilkes. 1998. "Thermal analysis of 1-ethyl-3-methylimidazolium tetrafluoroborate Molten Salt." *Proceedings of the Electrochemical Society* 98: 254–260. doi: 10.1149/199811.0254PV.

[56] Seddon, K. R., A. Stark, and M. J. Torres. 2000. "Influence of Chloride, Water, and Organic Solvents on the Physical Properties of Ionic Liquids." *Pure and Applied Chemistry* 72: 2275–2287. doi:10.1351/pac200072122275.

[57] Baker, G. A., S. N. Baker, S. Pandey, and F. V. Bright. 2005. "An Analytical View of Ionic Liquids." *The Analyst* 130: 800–808. doi:10.1039/b500865b.

[58] Sanders, J. R., E. H. Ward, and C. L. Hussey. 1986. "Aluminum Bromide-1-Methyl-3-Ethylimidazolium Bromide Ionic Liquids." *Journal of the Electrochemical Society* 133: 325–330. doi:10.1149/1.2108570.

[59] Okoturo, O. O., and T. J. Vandernoot. 2004. "Temperature Dependence of Viscosity for Room Temperature Ionic Liquids." *Journal of Electroanalytical Chemistry* 568: 167–181. doi:10.1016/j.jelechem.2003.12.050.

[60] Endres, F., and S. Z. El Abedin. 2006. "Air and Water Stable Ionic Liquids in Physical Chemistry." *Physical Chemistry Chemical Physics* 8: 2101–2116. doi:10.1039/b600519p.

[61] Najdanovic-Visak, V., J. M. S. S. Esperanca, L. P. N. Rebelo, M. N.da Ponte, H. J. R. Guedes, K. R. Seddon, and J. Szydlowski. 2002. "Phase Behaviour of Room Temperature Ionic Liquid Solutions: An Unusually Large Co-Solvent Effect in (Water + Ethanol)." *Physical Chemistry Chemical Physics* 4: 1701–1703. doi:10.1039/b201723g.

[62] Anthony, J. L., E. J. Maginn, and J. F. Brennecke. 2001. "Solution Thermodynamics of Imidazolium-Based Ionic Liquids and Water." *The Journal of Physical Chemistry B* 105: 10942–10949. doi:10.1021/jp0112368.

[63] Domańska, U., and A. Marciniak. 2004. "Solubility of Ionic Liquid [Emim][PF_6] in Alcohols." *The Journal of Physical Chemistry B* 108: 2376–2382. doi:10.1021/jp030582h.

[64] Lu, J., C. L. Liotta, and C. A. Eckert. 2003. "Spectroscopically Probing Microscopic Solvent Properties of Room-Temperature Ionic Liquids with the Addition of Carbon Dioxide." *The Journal of Physical Chemistry A* 107: 3995–4000. doi:10.1021/jp0224719.

[65] Katayanagi, H., K. Nishikawa, H. Shimozaki, K. Miki, P. Westh, and Y. Koga. 2004. "Mixing Schemes in Ionic Liquid—H_2O Systems: A Thermodynamic Study." *The Journal of Physical Chemistry B* 108: 19451–19457. doi:10.1021/jp0477607.

[66] Fredlake, C. P., M. J. Muldoon, S. N. V. K. Aki, T. Welton, and P. Christopher. 2004. "Solvent Strength of Ionic Liquid/CO_2 Mixtures." *Physical Chemistry Chemical Physics* 6: 3280–3285. doi:10.1039/b400815d.

[67] Kazarian, S. G., B. J. Briscoe, and T. Welton. 2000. "Combining Ionic Liquids and Supercritical Fluids: In Situ ATR-IR Study of CO_2 Dissolved in Two Ionic Liquids at High Pressures." *Chemical Communications* 0: 2047–2048. doi:10.1039/B005514J.

[68] Camarata, L., S. J. Kazarian, P. A. Salter, and T. Welton. 2001. "Molecular States of Water in Room Temperature Ionic Liquids." *Physical Chemistry Chemical Physics* 3: 5192–5200. doi:10.1039/b106900d.

[69] Cornils, B., and W. A. Herrmann. 1998. *Aqueous-Phase Organometallic Catalysis: Concepts and Applications.* Weinheim: Wiley-VCH.

[70] Giernoth, R. "Task-Specific Ionic Liquids." *Angewandte Chemie International Edition* 49: 2834–2839. doi:10.1002/anie.200905981.

[71] Mirjafari, A., L. N. Pham, J. R. McCabe, N. Mobarrez, E. A. Salter, A. Weirzbicki, K. N. West, R. E. Sykora, and J. H. Davis. 2013. "Building a Bridge Between Aprotic and Protic Ionic Liquids." *RSC Advances* 3: 337–340. doi:10.1039/c2ra22752e.

[72] Ding, J., and D. W. Armstrong. 2005. "Chiral Ionic Liquids: Synthesis and Applications." *Chirality* 17: 281–292. doi:10.1002/chir.20153.

Cyanine Dye Aggregation within Ionic Liquid ... 231

[73] Fukaya, Y., Y. Lizuka, K. Sekikawa, and H. Ohno. 2007. "Bio Ionic Liquids: Room Temperature Ionic Liquids Composed Wholly of Biomaterials." *Green Chemistry* 9: 1155. doi:10.1039/b706571j.

[74] Mehnert, C. P., R. A. Cook, N. C. Dispenziere, and M. J. Afeworki. 2002. "Supported Ionic Liquid Catalysis − A New Concept for Homogeneous Hydroformylation Catalysis." *Journal of the American Chemical Society* 124: 12932–12933. doi:10.1021/ja0279242.

[75] Karousos, D. S., E. Kouvelos, A. Sapalidis, K. Pohako-Esko, M. Bahlmann, P. S. Schulz, P. Wasserscheid, E. Siranidi, O. Vangeli, P. Falaras, N. Kanellopoulos, N. Romanos, and G. Em Romanos. 2016. "Novel Inverse Supported Ionic Liquid Absorbents for Acidic Gas Removal from Flue Gas." *Industrial & Engineering Chemistry Research* 55: 5748–5762. doi:10.1021/acs.iecr.6b00664.

[76] MacFarlane, D. R., J. M. Pringle, K. M. Johansson, S. A. Forsyth, and M. Forsyth. 2006. "Lewis base ionic liquids." *Chemical Communications* 0: 1905–1917. doi:10.1039/B516961P.

[77] Zanatta, M., A. Girard, N. M. Simon, G. Ebeling, H. K. Stassen, P. R. Livotto, F. P. dos Santos, and J. Dupont. 2014. "The Formation of Imidazolium Salt Intimate (Contact) Ion Pairs in Solution." *Angewandte Chemie International Edition* 53: 12817–12821. doi:10.1002/anie.201408151.

[78] Visser, A. E., R. P. Swatlowski, W. M. Reichert, R. Mayton, S. Sheff, A. Wierzbicki, J. H. Davis Jr., and R. D. Rogers. 2001. "Task-Specific Ionic Liquids for the Extraction of Metal Ions from Aqueous Solutions." *Chemical Communications* 0: 135–136. doi:10.1039/b008041l.

[79] Capello, C., U. Fischer, and K. Hungerbuhler. 2007. "What Is a Green Solvent? A Comprehensive Framework for the Environmental Assessment of Solvents." *Green Chemistry* 9: 927–934. doi:10.1039/b617536h.

[80] Clark, J. H., and S. J. Tavener. 2007. "Alternative Solvents: Shades of Green." *Organic Process Research & Development* 11: 149–155. doi:10.1021/op060160g.

[81] Zhang, X., C. Li, C. Fu, and S. Zhang. 2008. "Environmental Impact Assessment of Chemical Process Using the Green Degree Method." *Industrial & Engineering Chemistry Research* 47: 1085–1094. doi:10.1021/ie0705599.

[82] Dabiri, M., M. Baghbanzadeh, M. S. Nikcheh, and E. Arzroomchilar. 2008. "Eco Friendly and Efficient One-Pot Synthesis of Alkyl- or Aryl-14H-Dibenzo[a,j]Xanthenes in Water." *Bioorganic & Medicinal Chemistry Letters* 18: 436–438. doi:10.1016/lj.bmcl.2007. 07.008.

[83] Lu, L., K. Ai, and Y. Ozaki. 2008. "Environmentally Friendly Synthesis of Highly Monodisperse Biocompatible Gold Nanoparticles with Urchin-like Shape." *Langmuir* 24: 1058–1063. doi:10.1021/la702886q.

[84] Ion, A., C. Van Doorslaer, V. Parvulescu, P. Jacobs, and D. De Vos. 2008. "Green Synthesis of Carbamates from CO_2, Amines and Alcohols." *Green Chemistry* 10: 111–116. doi:10.1039/B711197.

[85] Poliakoff, M., and P. Licence. 2007. "Green Chemistry." *Nature* 450: 810–812. doi:10.1038/450810a.

[86] Smith, N. M., C. L. Raston, C. B. Smith, and A. N. Sobolev. 2007. "PEG Mediated Synthesis of Amino-Functionalised 2,4,6-Triarylpyridines." *Green Chemistry* 9: 1185–1190. doi:10.1039/b700893g.

[87] Lei, H., A. Pizzi, and G. Du. 2007. "Environmentally Friendly Mixed Tannin/Lignin Wood Resins." *Journal of Applied Polymer Science* 107: 203–209. doi:10.1002/app.27011.

[88] Abbott, A. P., G. Capper, D. L. Davies, R. K. Rasheed, and V. Tambyrajah. 2003. "Novel Solvent Properties of Choline Chloride/Urea." *Chemical Communications* 0: 70–71. doi:10.1039/b210714g.

[89] Abbott, A. P., D. Boothby, G. Capper, D. L. Davies, and R. K. Rasheed. 2004. "Deep Eutectic Solvents Formed between Choline Chloride and Carboxylic Acids: Versatile Alternatives to Ionic Liquids." *Journal of the American Chemical Society* 126: 9142–9147. doi:10.1021/ja048266j.

[90] Dai, Y., J. V. Spronsen, G. J. Witkamp, R. Verpoorte, and Y. H. Choi. 2013. "Natural Deep Eutectic Solvents as New Potential Media for Green Technology." *Analytica Chimica Acta* 766: 61–68. doi:10.1016/j.aca.2012.12.019.

[91] Abbott, A. P., G. Capper, D. L. Davies, R. K. Rasheed, and V. Tambyrajah. 2002. "Quaternary Ammonium Zinc- or Tin-Containing Ionic Liquids: Water Insensitive, Recyclable Catalysts for Diels–Alder Reactions." *Green Chemistry* 4: 24–26. doi:10.1039/b108431c.

[92] Zhang, Q., K. D. O. Vigier, S. Royer, and F. Jerome. 2012. "Deep Eutectic Solvents: Syntheses, Properties and Applications." *Chemical Society Reviews* 41: 7108–7146. doi:10.1039/c2cs35178a.

[93] Pandey, A., Bhawna, D. Dhingra, and S. Pandey. 2017. "Hydrogen Bond Donor/Acceptor Cosolvent-Modified Choline Chloride-Based Deep Eutectic Solvents." *The Journal of Physical Chemistry B* 121: 4202–4212. doi:10.1021/acs.jpcb.7b01724.

[94] Abbott, A. P., C. A. Eardley, N. R. S. Farley, G. A. Griffith, and A. Pratt. 2001. "Electrodeposition of Aluminium and Aluminium/ Platinum Alloys from AlCl$_3$/Benzyltrimethyl-ammonium Chloride Room Temperature Ionic Liquids." *Journal of Appplied Electrochemistry 31*: 1345–1350. https://link.springer.com/content/pdf/10.1023%2FA%3A1013800721923.pdf.

[95] Abbott, A. P., G. Capper, K. J. McKenzie, and K. S. Ryder. 2007. "Electrodeposition of Zinc–Tin Alloys from Deep Eutectic Solvents Based on Choline Chloride." *Journal of Electroanalytical Chemistry* 599: 288–294. doi:10.1016/j.jelechem.2006.04.024.

[96] Smith, E. L., A. P. Abbott, and K. S. Ryder. 2014. "Deep Eutectic Solvents (DESs) and Their Applications." *Chemical Reviews* 114: 11060–11082. doi:10.1021/cr300162p.

[97] Choi, Y. H., J. van Spronsen, Y. T. Dai, M. Verberne, F. Hollmann, I. Arends, G. J. Witkamp, and R. Verpoorte. 2011. "Are Natural Deep Eutectic Solvents the Missing Link in Understanding Cellular Metabolism and Physiology." *Plant Physiology* 156: 1701–1705. doi.org/10.1104/pp.111.178426.

[98] D. van Osch, L. F. Zubeir, A. V. D. Bruinhorst, M. A. A. Rocha, and M. C. Kroon. 2015. "Hydrophobic deep eutectic solvents as water-immiscible extractants." *Green Chemistry* 17: 4518–4521. doi:10.1039/C5GC01451D.

[99] Hsiu, S. I., J. F. Huang, and I. W. Sun. 2002. "Lewis Acidity Dependency of the Electrochemical Window of Zinc Chloride-1-Ethyl-3-Methylimidazolium Chloride Ionic Liquids." *Electrochimica Acta* 47: 4367–4372. doi:10.1016/s0013-4686(02) 00509-1.

[100] Lin, Y. F., and I. W. Sun. 1999. "Electrodeposition of Zinc from a Lewis Acidic Zinc Chloride-1-Ethyl-3-Methylimidazolium Chloride Molten Salt." *Electrochimica Acta* 44: 2771–2777. doi:10.1016/s0013-4686(99)00003-1.

[101] Sitze, M. S., E. R. Schreiter, E. V. Patterson, and R. G. Freeman. 2001. "Ionic Liquids Based on $FeCl_3$ and $FeCl_2$. Raman Scattering and Ab Initio Calculations." *Inorganic Chemistry* 40: 2298–2304. doi:10.1021/ic001042r.

[102] Scheffler, T. B., and M. S. Thomson. 1990. *Seventh International Conference on Molten Salts*; The Electrochemical Society: Montreal, 281.

[103] Abbott, A. P., G. Capper, D. L. Davies, R. K. Rasheed, J. Archer, and C. John. 2004. "Electrodeposition of Chromium Black from Ionic Liquids." *Transactions of the IMF* 82: 14–17. doi:10.1080/00202967.2004.11871547.

[104] Yang, J. Z., P. Tian, L. L. He, and W. G. Xu. 2003. "Studies on Room Temperature Ionic Liquid $InCl_3$–EMIC." *Fluid Phase Equilibria* 204: 295–302. doi:10.1016/s0378-3812(02)00265-0.

[105] Francisco, M., A. V. D. Bruinhorst, and M. C. Kroon. 2013. "Low-Transition-Temperature Mixtures (LTTMs): A New Generation of Designer Solvents." *Angewandte Chemie International Edition* 52: 3074–3085. doi:10.1002/anie.201207548.

[106] Ruß, C., and B. König. 2012. "Low Melting Mixtures in Organic Synthesis – an Alternative to Ionic Liquids?" *Green Chemistry* 14: 2969–2982. doi:10.1039/c2gc36005e.

[107] Obeten, M. E., B. U. Ugi, and N. O. Alobi. 2017. "A Review on Electrochemical Properties of Choline Chloride Based Eutectic Solvent in Mineral Processing." *Journal of Applied Sciences and Environmental Management* 21: 991–998. doi:10.4314/jasem.v21i5.29.

[108] Gorke, J. T., and R. J. Kazlauskas. 2008. "Hydrolase-Catalyzed Biotransformations in Deep Eutectic Solvents." *Chemical Communications* 0: 1235–1237. doi:10.1039/b716317g.

[109] Manurung, R., A. Syahputra, and M. A. Alhamd. 2018. "Purification of Palm Biodiesel Using Deep Eutectic Solvent (DES) Based Choline Chloride (ChCl) and 1,2-Propanediol ($C_3H_8O_2$)." *Journal of Physics: Conference Series* 1028: 012032. doi:10.1088/1742-6596/1028/1/012032.

[110] Dhingra, D., Bhawna, A. Pandey, and S. Pandey. 2019. "Pyrene Fluorescence to Probe a Lithium Chloride-Added (Choline Chloride + Urea) Deep Eutectic Solvent." *The Journal of Physical Chemistry B* 123: 3103–3111. doi:10.1021/acs.jpcb.9b01193.

[111] Weaver, K. D., H. J. Kim, J. Sun, D. R. MacFarlane, and G. D. Elliott. 2010. "Cyto-Toxicity and Biocompatibility of a Family of Choline Phosphate Ionic Liquids Designed for Pharmaceutical Applications." *Green Chemistry* 12: 507–513. doi:10.1039/b918726j.

[112] Zhao, H., and G. A. Baker. 2013. "Ionic Liquids and Deep Eutectic Solvents for Biodiesel Synthesis: A Review" *Journal of Chemical Technology and Biotechnology* 88: 3–12. doi:10.1002/jctb.3935.

[113] Gutiérrez-Arnillas, E., M. S. Alvarez, F. J. Deive, A. Rodriguez, and M. A. Sanroman. 2016. "New Horizons in the Enzymatic Production of Biodiesel Using Neoteric Solvents." *Renewable Energy* 98: 92–100. doi:10.1016/j.renene.2016.02.058.

[114] Zhao, H., C. Zhang, and T. D. Crittle. 2013. "Choline-Based Deep Eutectic Solvents for Enzymatic Preparation of Biodiesel from Soybean Oil." *Journal of Molecular Catalysis B: Enzymatic* 85: 243–247. doi:10.1016/j.molcatb.2012.09.003.

[115] Domínguez De M. P., and Z. Maugeri. 2011. "Ionic Liquids in Biotransformations: from Proof-of-Concept to Emerging Deep-Eutectic-Solvents." *Current Opinion in Chemical Biology* 15: 220–225. doi:10.1016/j.cbpa.2010.11.008.

[116] Bhawna, A. Pandey, and S. Pandey. 2017. "Superbase-Added Choline Chloride-Based Deep Eutectic Solvents for CO_2 Capture and Sequestration." *ChemistrySelect* 2: 11422–11430. doi:10.1002/slct.201702259.

[117] Wagle, D. V., H. Zhao, and G. A. Baker. 2014. "Deep Eutectic Solvents: Sustainable Media for Nanoscale and Functional Materials." *Accounts of Chemical Research* 47: 2299–2308. doi:10.1021/ar5000488.

[118] Carriazo, D., M. C. Serrano, M. C. Gutierrez, M. L. Ferrer, and F. del Monte. 2012. "Deep-Eutectic Solvents Playing Multiple Roles in the Synthesis of Polymers and Related Materials." *Chemical Society Reviews* 41: 4996–5014. doi:10.1039/c2cs15353j.

[119] Atilhan, M., L. T. Costa, and S. Aparicio. 2017. "Elucidating the Properties of Graphene-Deep Eutectic Solvents Interface." *Langmuir* 33: 5154–5165. doi:10.1021/acs.langmuir.7b00767.

[120] Hosu, O., M. M. Barsan, C. Cristea, R. Sandulescu, and C. M. A. Brett. 2017. "Nanostructured Electropolymerized Poly(Methylene Blue) Films from Deep Eutectic Solvents. Optimization and Characterization." *Electrochimica Acta* 232: 285–295. doi:10.1016/j.electacta.2017.02.142.

[121] Daehne, S. 1978. "Color and Constitution: One Hundred Years of Research." *Science* 199: 1163–1167. doi: 10.1126/science. 199.4334.1163.

[122] Zhang, L., and J. M. Cole. 2017. "Dye Aggregation in Dye-Sensitized Solar Cells." *Journal of Materials Chemistry A* 5: 19541–19559. doi:10.1039/c7ta05632j.

[123] Takahashi, D., H. Oda, T. Izumi, and R. Hirohashi. 2005. "Substituent Effects on Aggregation Phenomena in Aqueous Solution of Thiacarbocyanine Dyes." *Dyes and Pigments* 66: 1–6. doi: 10.1016/j.dyepig.2004.08.008.

[124] Honda, C., and H. Hada. 1977. "Spectroscopic Study on the J-Aggregate of Cyanine Dyes: Effect of Substituents on Molecular Association of Cyanine Dyes." *Photographic Science and Engineering* 21: 97–102. https://www.researchgate.net/publication/279589459_SPECTROSCOPIC_STUDY_ON_THE_J-AGGREGATE_OF_CYANINE_DYES_-_3.

[125] Mishra, A., R. K. Behera, P. K. Behera, B. K. Mishra, and G. B. Behera. 2000. "Cyanines During the 1990s: A Review" *Chemical Reviews* 100: 1973–2012. doi:10.1021/cr990402t.

[126] Kobayashi, T. 1996. *J Aggregates.* London: World Scientific Publishing.

[127] McRae, E. G., and M. Kasha. 1958. "Enhancement of Phosphorescence Ability upon Aggregation of Dye Molecules." *The Journal of Chemical Physics* 28: 721–722. doi:10.1063/1.1744225.

[128] Kasha, M., H. R. Rawis, and M. A. El-Bayoumi. 1965. "The Exciton Model in Molecular Spectroscopy." *Pure and Applied Chemistry* 11: 371–392. doi:10.1351/pac196511030371.

[129] Franck, J., and E. Teller. 1938. "Migration and Photochemical Action of Excitation Energy in Crystals." *The Journal of Chemical Physics* 6: 861–872. doi:10.1063/1.1750182.

[130] Burshtein, K. Ya. 1987. "Mo Studies of Solvation Effects." *Journal of Molecular Structure: THEOCHEM* 153: 209–213. doi:10.1016/0166-1280(87)80003-9.

[131] Gudipati, M. S. 1994. "Exciton, Exchange, and Through-Bond Interactions in Multichromophoric Molecules: An Analysis of the Electronic Excited States." *The Journal of Physical Chemistry* 98: 9750–9763. doi:10.1021/j100090a007.

[132] Kasha, M. 1963. "Energy Transfer Mechanisms and the Molecular Exciton Model for Molecular Aggregates." *Radiation Research* 20: 55–71. doi:10.2307/3571331.

[133] Harrison, W. J., D. L. Mateer, and G. J. T. Tiddy, 1996. "Liquid-Crystalline J-Aggregates Formed by Aqueous Ionic Cyanine Dyes." *The Journal of Physical Chemistry* 100: 2310–2321. doi:10.1021/jp9525321.

[134] Herz, A. H. 1977. "Aggregation of Sensitizing Dyes in Solution and Their Adsorption onto Silver Halides." *Advances in Colloid and Interface Science* 8: 237–298. doi:10.1016/0001-8686(77)80011-0.

[135] Borsenberger, P. M., A. Chowdry, D. Hoesterey, and W. Mey. 1978. "An Aggregate Organic Photoconductor. II. Photoconduction Properties." *Journal of Applied Physics* 49: 5555–5564. doi:10.1063/1.324476.

[136] Waggoner, A. 1976. "Optical Probes of Membrane Potential." *The Journal of Membrane Biology* 27: 317–334. doi:10.1007/bf01869143.

[137] Slavnova, T. D., A. K. Chibisov, and H. Gorner. 2005. "Kinetics of Salt-Induced J-Aggregation of Cyanine Dyes." *The Journal of Physical Chemistry A*: 4758–4765. doi:10.1021/jp058014k.

[138] Hanamura, E. 1988. "Very Large Optical Nonlinearity of Semiconductor Microcrystallites." *Physical Review B* 37: 1273–1279. doi:10.1103/physrevb.37.1273.

[139] Koti, A. S. R., J. Taneja, and N. Periasamy. 2003. "Control of Coherence Length and Aggregate Size in the J-Aggregate of Porphyrin." *Chemical Physics Letters* 375: 171–176. doi:10.1016/S0009-2614(03)00866-2.

[140] Xiang, J., X. Yang, C. Chen, Y. Tang, W. Yan, and G. Xu. 2003. "Effects of NaCl on the J-Aggregation of Two Thiacarbocyanine Dyes in Aqueous Solutions." *Journal of Colloid and Interface Science* 258: 198–205. doi:10.1016/s0021-9797(02)00187-x.

[141] Chibisov, A. K., H. Gorner, and T. D. Slavnova. 2004. "Kinetics of Salt-Induced J-Aggregation of an Anionic Thiacarbocyanine Dye in Aqueous Solution." *Chemical Physics Letters* 390: 240–245. doi:10.1016/j.cplett.2004.03.131.

[142] Zhang, Y., J. F. Xiang, Y. L. Tang, G. Z. Xu, and W. P. Yan. 2008. "Aggregation Behaviour of Two Thiacarbocyanine Dyes in Aqueous Solution." *Dyes and Pigments* 76: 88–93. doi:10.1016/j.dyepig.2006.08.009.

[143] Kobayashi, T. 2012. *J-Aggregates Volume 2.* Japan: World Scientific.

[144] West, W., and S. J. Pearce. 1965. "The Dimeric State of Cyanine Dyes." *The Journal of Physical Chemistry* 69: 1894–1903. doi:10.1021/j100890a019.

[145] Sasaki, F., and S. Kobayashi. 1993. "Anomalous Excitation Density Dependence of Nonlinear Optical Responses in Pseudoisocyanine J Aggregates." *Applied Physics Letters* 63: 2887–2889. doi:10.1063/1.110315.

[146] Wang, Y. 1986. "Efficient Second-Harmonic Generation from Low-Dimensional Dye Aggregates in Thin Polymer Film." *Chemical Physics Letters* 126: 209–214. doi:10.1016/s0009-2614(86)80041-0.

[147] Wang, Y. 1991. "Resonant Third-Order Optical Nonlinearity of Molecular Aggregates with Low-Dimensional Excitons." *Journal of the Optical Society of America B* 8: 981–985. doi:10.1364/josab.8.000981.

[148] Dolphin, D. 1978. *The Porphyrins V3 1st Edition Physical Chemistry, Part A* Canada: Academic Press. doi: 10.1016/B978-0-12-220103-5.X5001-6.

[149] Behera, K., and S. Pandey. 2008. "Ionic Liquid Induced Changes in the Properties of Aqueous Zwitterionic Surfactant Solution." *Langmuir* 24: 6462–6469. doi:10.1021/la800141p.

[150] Kumar, V., G. A. Baker, and S. Pandey. 2011. "Ionic Liquid-Controlled J- versus H-aggregation of Cyanine Dyes." *Chemical Communications* 47: 4730–4732. doi:10.1039/c1cc00080b.

[151] Kumar, V., G. A. Baker, S. Pandey, S. N. Baker, and S. Pandey. 2011. "Contrasting Behavior of Classical Salts versus Ionic Liquids toward Aqueous Phase J-Aggregate Dissociation of a Cyanine Dye." *Langmuir* 27: 12884–12890. doi.org/10.1021/la203317t.

[152] Pal, P., A. Yadav, and S. Pandey. 2017. "Aggregation of Carbocyanine Dyes in Choline Chloride-Based Deep Eutectic Solvents in the Presence of an Aqueous Base." *Langmuir* 33: 9781–9792. doi:10.1021/acs.langmuir.7b01981.

INDEX

A

absorption spectra, 21, 56, 59, 61, 63, 64, 66, 73, 76, 77, 79, 80, 83, 84, 85, 89, 92, 97, 98, 105, 120, 130, 132, 139, 144, 145, 162, 163, 165, 182, 188, 204, 206, 212, 215
absorption spectroscopy, 105, 209
access, 72, 137, 223
accounting, 65, 157
acid, 5, 10, 39, 116
acidic, 47, 197
additives, 207, 213, 221
adsorption, 120, 158, 161, 166, 168, 169, 175, 203
aggregation process, 163, 208, 209, 222
albumin, 3, 60, 82, 86, 87, 97
algorithm, 11, 60
amino, 19, 31, 35, 37, 57, 87, 102, 103, 185, 200, 203
amino acid, 19, 31, 35, 37, 57, 87, 102, 103, 200, 203
ammonium, 195, 201, 208, 210, 233
amyloid fibril formation, vii, viii, 2, 12, 22, 29, 39, 44, 46, 51, 57, 87
amyloid fibrils, 2, 3, 11, 32, 38, 41, 46, 47, 54, 58, 87, 97, 106, 120, 121

amyloidosis, viii, 2, 3, 43, 45, 50, 87
anisotropy, 187, 191
atomic force, 47, 135, 139
atoms, 78, 130, 166, 168, 169, 175
Au nanoparticles, 140, 144, 151, 159

B

bacteria, 3, 124
base pair, 59, 67, 69, 71, 72
benign, 194, 198, 199, 203
binding energy, 137, 175
biodegradability, 194, 201
biomass, 201, 203
biomedical applications, 3, 40, 203
biomolecules, x, 40, 54, 55, 56, 109, 123, 125, 174
biopolymers, 55, 57
biosynthesis, 44, 180
biotechnology, ix, 2, 134
blocking the lateral extension of β-sheets, viii, 2, 35, 39
blood, 3, 4, 26, 39, 47
blood-brain barrier, 4, 26, 39, 47
bonding, 38, 195, 197, 199
bovine serum albumin, 57, 60, 82, 83, 86, 87, 97, 106, 109, 118

C

cancer, 3, 134, 173, 206
capillary, 3, 40, 118
carbocyanine dyes, 239
carbon, 10, 125
catalysis, 134, 141, 174, 195, 198
cation, 195, 197, 200
chaos, 164, 179
chemical, viii, 2, 4, 5, 10, 26, 27, 28, 29, 30, 48, 64, 65, 133, 134, 137, 141, 163, 171, 172, 174, 194, 196, 202, 206
chemical characteristics, 5, 29
chemical properties, 194, 202
chemical stability, 134, 172
choline, 78, 199, 201
clusters, 166, 187
color, iv, 107, 115, 204
community, 200, 204
competition, 21, 94
composition, 144, 172
compounds, viii, 2, 3, 4, 5, 8, 11, 13, 14, 15, 16, 17, 18, 20, 21, 22, 25, 26, 30, 31, 32, 33, 34, 36, 38, 39, 55, 64, 125, 129, 174, 208
conduction, 136, 170
conductivity, 137, 138, 139, 184, 202
construction, 134, 170
correlation, 5, 26, 28, 29, 35, 39, 120, 138
correlation coefficient, 27, 29
corrosion, 55, 107
cost, 136, 170, 201, 202, 220
covering, 129, 142, 145, 174
crystal structure, 11, 86
CTAB, 139, 142, 219, 221
curcumin, 4, 29
cyanine compounds, 2, 4, 5, 13, 14, 17, 18, 21, 22, 25, 26, 30, 31, 37, 39
cyanine-DNA binding, 71
cytometry, 40, 114
cytotoxicity, 4, 39, 41, 49, 134

D

deep eutectic solvents, vii, x, 193, 194, 199, 232, 233, 234, 235, 236, 239
density functional theory, 162, 166
derivatives, 3, 4, 41, 46, 48, 57, 61, 64, 67, 72, 111, 124
destruction, 57, 176
detection, x, 3, 46, 49, 54, 55, 67, 68, 82, 87, 105, 109, 110, 114, 115, 116, 120, 123, 125, 136, 171, 172, 174, 185
detection techniques, x, 123, 174
DFT calculations, 124, 141, 174
dielectric constant, 80, 129, 207, 214
dimerization, 76, 88
dipole moments, 30, 163
dipoles, 56, 131, 187, 205
diseases, 3, 46
disorder, 147, 163, 185
dispersion, 55, 65, 76, 79, 92, 126, 139, 143, 159, 173, 197
displacement, 22, 118
dissociation, 77, 85, 104
distribution, 135, 137, 140, 141, 150, 151, 166, 197
distribution function, 150
DMTC, 210, 212, 216, 217, 218, 219, 221
DNA, ix, 3, 4, 39, 40, 42, 43, 45, 48, 54, 57, 58, 59, 67, 68, 69, 70, 71, 72, 73, 74, 75, 83, 101, 102, 103, 104, 105, 109, 111, 112, 113, 114, 115, 116, 119, 120, 171, 180
double stranded DNA, 59, 67, 74, 106
drug delivery, vii, 2, 3, 141, 174, 203
drugs, 3, 4, 87, 118, 203
dye monomer-aggregate equilibria, 88
dye monomers, 11, 21, 64, 65, 74, 80, 81, 83, 84, 88, 89, 90, 91, 94, 95, 96, 100, 106, 207, 214

Index 243

E

egg, 57, 60
electric charge, 86, 134
electrochemistry, 133, 198, 203
electroluminescence, 130, 170
electromagnetic, 137, 166
electron, viii, 2, 5, 6, 10, 22, 24, 55, 108, 124, 126, 127, 131, 135, 136, 137, 162, 164, 170
electron microscopy, viii, 2, 6, 10, 24
electronic structure, 44, 209
electrophoresis, 3, 40, 67, 118
elongation, viii, 2, 19, 22, 23, 32, 38, 39
elucidation, 46, 153
emission, ix, x, 3, 9, 21, 49, 54, 55, 68, 69, 106, 119, 123, 125, 130, 137, 143, 147, 148, 149, 169, 171, 172, 174, 204, 209, 212, 214, 215, 217, 218, 220
emitters, 173
energy, 10, 21, 32, 55, 56, 60, 68, 75, 124, 127, 131, 137, 144, 147, 154, 162, 163, 166, 167, 169, 170, 176, 177, 185
energy transfer, 21, 32, 124, 144, 154
environment, 67, 74, 75, 76, 78, 79, 81, 105, 131, 162, 169, 173, 175, 194, 208
environmental conditions, 56
environmental factors, 19, 46
equilibrium, 47, 88, 89, 96, 104, 155, 159, 161, 163, 171
evidence, viii, 2, 22, 23, 38, 159
excitation, 9, 21, 59, 82, 124, 131, 146, 148, 162, 173, 186
exciton, 56, 110, 127, 129, 131, 132, 134, 141, 144, 145, 148, 163, 165, 166, 170, 173, 179, 187, 204
exclusion, 59, 70, 72
experimental condition, 68, 154, 158, 160, 165, 172
extinction, 3, 6, 8, 58, 62, 88, 91, 95, 128, 130, 133, 165, 166, 204

F

fibrillation, 41, 43, 49, 50, 118
filament, 41, 47
films, 55, 110
flammability, 194, 195, 202
flexibility, 19, 37, 72
fluctuations, 12, 137, 146
fluorescence, viii, ix, x, 2, 3, 5, 9, 10, 11, 12, 13, 15, 20, 21, 22, 23, 25, 32, 39, 40, 42, 43, 47, 54, 55, 57, 59, 67, 68, 70, 82, 87, 106, 108, 109, 110, 114, 115, 121, 124, 125, 128, 130, 131, 141, 142, 143, 146, 147, 153, 154, 155, 157, 158, 159, 165, 168, 170,171, 172, 173, 174, 175, 176, 177, 178, 179, 181, 182, 183, 184, 186, 188, 190, 191, 204, 209, 210, 211, 212, 214, 215, 217, 218, 219, 220, 235
fluorophores, ix, 54, 72, 78, 108, 119
force, 12, 45, 76, 162, 163, 168
formation, vii, viii, x, 2, 4, 5, 6, 9, 11, 12, 16, 19, 21, 22, 27, 29, 32, 33, 34, 36, 38, 39, 42, 44, 45, 46, 47, 49, 50, 51, 55, 56, 71, 75, 76, 79, 81, 82, 83, 84, 85, 87, 88, 89, 94, 96, 97, 101, 103, 106, 108, 111, 113, 115, 119, 120, 124, 129, 130, 132, 134,136, 142, 145, 153, 154, 158, 159, 160, 161, 170, 171, 172, 173, 174, 176, 195, 197, 201, 204, 206, 208, 210, 212, 217, 219, 221

G

geometry, 10, 56, 60, 131, 135, 167, 169, 204, 205, 209
glyceline, 217, 218, 219, 220, 221
glycerol, 57, 217
glycine, 8, 58
gold nanoparticles, 158, 190
growth, 9, 13, 29, 31, 57, 74, 105, 159, 175, 215

H

H- aggregates, 127
H- and J-aggregates, 113, 194, 204, 206
H-aggregates, x, 56, 65, 75, 76, 77, 79, 81, 82, 84, 85, 88, 89, 90, 91, 93, 94, 95, 96, 97, 99, 100, 101, 103, 104, 105, 120, 124, 130, 131, 163, 164, 174, 179, 205, 208, 210, 212, 218, 219, 221
harvesting, 110, 124, 125, 129, 148, 170, 177, 181
H-bonding, 66, 94, 216
height, 11, 140, 166
helicity, 35, 36, 75
histogram, 140, 150
human, 3, 11, 30, 34, 35, 45, 47, 49, 50, 57, 82, 94, 97, 109, 111, 112, 113, 117, 118, 194
hybrid, 125, 126, 134, 141, 162, 166, 170
hydrogen, 72, 195, 197, 199, 208

I

identity, x, 194, 197, 208, 210, 212, 221
in vitro, viii, 2, 4, 5, 39, 45, 47, 49, 105, 114
in vivo, 4, 5, 40, 42, 172, 173
infrared spectroscopy, 135, 174
inhibition, 15, 21, 22, 31, 35, 38, 41, 42, 44, 46, 47, 49, 50, 51, 112, 118
inhibitor, 9, 19, 21, 48
inhibitory potency, viii, 2, 5, 16, 17, 26, 29, 30, 34, 36, 39
insulin, v, vii, viii, 1, 2, 4, 5, 6, 8, 9, 10, 11, 12, 13, 14, 15, 17, 18, 19, 20, 21, 22, 24, 25, 26, 29, 30, 31, 32, 33, 34, 35, 36, 37, 38, 39, 43, 44, 46, 47, 49, 50, 51, 94, 97, 112
insulin monomer, viii, 2, 12, 33
interface, viii, 2, 32, 78, 104, 124, 169

ionic liquids, vii, x, 193, 194, 195, 197, 209, 213, 224, 225, 226, 227, 228, 229, 230, 231, 232, 233, 234, 235, 236, 239
ions, 12, 80, 134, 137, 138, 163, 168, 195, 197, 208, 209
irradiation, 44, 137
isomerization, 55, 82
isotherms, 69, 70

J

J- aggregates, 130, 131, 171
J- and H-aggregation, 124, 174, 175

L

labeling, vii, ix, 2, 3, 40, 43, 54, 55, 57, 106, 108, 109
lateral extension, 2, 32
lifetime, 111, 217
ligand, 11, 59, 60, 86, 153, 155, 156
light, 55, 107, 110, 124, 125, 129, 130, 135, 136, 137, 138, 146, 148, 153, 164, 165, 170, 172, 173, 177, 181, 184, 204, 209
light scattering, 135, 137, 138, 172, 209
lipids, 4, 41, 55, 57, 78, 172, 173
liposomes, 54, 58, 59, 76, 77, 80, 81, 82, 105, 109
liquids, vii, x, 193, 194, 197, 198, 231
luminescence, 111, 115
lysozyme, ix, 41, 45, 50, 54, 57, 58, 59, 60, 88, 92, 93, 94, 95, 96, 97, 98, 99, 100, 101, 102, 103, 104, 105, 106, 112, 113, 121

M

macromolecules, vii, ix, x, 54, 57, 71, 124, 129, 132, 137, 142, 153, 155, 171, 174 magnitude, 21, 36, 68, 70, 73, 77, 79, 103, 106, 134

media, ix, x, 54, 65, 109, 193, 194, 195, 198, 200, 203, 207, 209, 214, 216, 217, 219, 222

medical, 141, 174

medicine, ix, x, 2, 55, 175, 191, 193

melting, 195, 197, 199

membranes, ix, 54, 59, 78, 79, 82, 116, 172

metal ion, x, 124, 129, 158, 161, 198, 208

metal nanoparticles, 136, 183

methanol, 5, 8, 58, 61, 62, 63, 119, 188

methylene blue, 115, 116

microscope, 10, 130

microscopy, 5, 9, 22, 25, 39, 40, 47, 67, 114, 135, 139, 146, 172, 173

migration, 124, 127, 138, 170, 173

mixing, 160, 201, 203, 217

molecular docking studies, viii, 2, 11, 31

molecular dynamics simulations, viii, 2, 5, 12, 22, 31, 35, 39, 48, 180

molecular weight, 4, 11, 39

molecules, vii, x, 1, 3, 4, 12, 21, 22, 30, 34, 36, 38, 43, 45, 50, 55, 57, 60, 67, 71, 75, 76, 85, 87, 92, 110, 112, 113, 114, 116, 124, 125, 126, 127, 129, 130, 131, 132, 134, 136, 137, 139, 141, 147, 148, 149, 151, 152, 153, 155, 156, 158, 159, 161, 162, 163, 166, 169, 174, 178, 185, 186, 187, 200, 204, 208, 209, 216

monolayer, 110, 183

monomers, 11, 12, 21, 22, 30, 33, 64, 65, 74, 76, 79, 80, 81, 82, 83, 84, 85, 88, 89, 90, 91, 93, 94, 95, 96, 99, 100, 103, 104, 106, 119, 129, 130, 142, 144, 159, 163, 164, 166, 207, 208, 213, 214, 216, 218, 219, 221

morphology, 135, 138, 139, 141

N

nanoelectronics, 141, 174

nanomaterials, 133, 136, 137, 139, 198

nanoparticles, viii, x, 2, 124, 128, 133, 140, 144, 158, 159, 164, 166, 172, 173, 174, 176, 180, 182, 183, 184, 186, 187, 188, 189, 190, 191

nanorods, 139, 144

nanospectroscopy methods, 124

nanosystems, 148, 165

nanotechnology, x, 134, 193

native and fibrillar lysozyme, 45, 58, 87, 92, 105, 106, 112

neurodegenerative diseases, 41, 42

neutral, 78, 131, 213

nitrogen, vii, ix, 1, 5, 16, 17, 19, 39, 53, 54, 58, 125

nucleation, viii, 2, 12, 16, 19, 22, 23, 34, 38, 39, 182

nucleic acid, vii, ix, x, 2, 3, 43, 54, 55, 57, 67, 71, 74, 105, 108, 113, 114, 115, 116, 121, 123, 125, 174

nucleus, 12, 33, 35

O

oligomerization, 47, 81, 82

oligomers, 23, 42, 49, 50, 110, 144

optical properties, x, 124, 131, 133, 134, 136, 162, 166, 172, 174

optimization, 10, 114, 167, 169

organic compounds, x, 123

organic solvents, 199, 203

oxygen, 154, 166, 168

P

parallel, x, 30, 31, 55, 76, 124, 127, 131, 174, 205, 215

peptides, x, 34, 41, 60, 4, 94, 104, 118, 123, 125, 174

permission, iv, 211, 212, 213, 214, 215, 218, 220

pharmaceutical, vii, viii, 2, 4, 39, 114

pharmacology, ix, 3, 54, 55, 180

phosphate, 58, 59, 78

phosphatidylcholine, 105, 108, 117

photodynamic therapy, vii, 2, 3, 40, 108, 206

photopolymerization, 55, 107

photosensitizers, x, 3, 123

photosynthetic systems, 169, 206

physical properties, x, 133, 134, 193, 197, 202

physicochemical properties, x, 78, 193, 195, 196, 209

polar, 3, 11, 31, 34, 44, 48, 76, 79, 87, 103, 172

polarity, 56, 65, 67, 69, 76, 79, 81, 96, 97, 116, 203, 207

polarizability, 11, 26, 55, 126, 127, 166, 217

polarization, 141, 144, 153, 177, 191

polydispersity, 138, 140

polymer, 109, 130, 165, 171, 173

polymer films, 109, 130, 173

polymerization, 48, 111, 182, 198

polymers, 129, 170, 177, 207

polypeptide, 41, 42, 49

polyphenols, 29, 39, 51

polysaccharides, 171, 203

prevention, 19, 27, 29, 32

probe, 49, 74, 82, 88, 89, 116, 117, 118

protein structure, viii, 2, 11, 35, 36

proteins, vii, ix, x, 2, 3, 25, 27, 47, 54, 55, 57, 60, 71, 105, 108, 120, 123, 124, 125, 171, 173, 174, 203

purification, 6, 58, 67, 203

Q

quantum-chemical calculations, 5, 22, 26, 30

quartz, 9, 59

quaternary ammonium, 200, 201

R

Radiation, 18, 2171, 237

radius, 12, 36, 37, 38, 135, 137, 165, 166

radius of gyration, 12, 36, 37, 38

Raman spectroscopy, 135, 141, 153, 182

reactions, 49, 107, 158, 161, 187

recognition, 73, 172

red shift, 56, 61, 67, 97, 130, 152, 164

relaxation, 117, 119, 160, 181, 187

reline, 217, 218, 219, 220, 221

researchers, 197, 200, 201, 202, 208, 209, 216

resveratrol, 19, 31

rings, vii, 1, 66, 131

rods, 129, 142

room temperature, 58, 79, 195

S

salts, 110, 132, 195, 201, 203, 207, 208, 213, 214, 216

saturation, 9, 13, 82, 153, 156

scattering, 116, 133, 134, 153, 164, 165, 170, 182

SDS, 219, 220, 221

selectivity, 22, 172, 195

selenium, 49, 110

self-assembly, viii, 2, 4, 23, 47, 74, 130, 134, 203

self-organization, vii, x, 56, 110, 124, 126, 127, 129, 130, 169, 174, 175, 206

semiconductor, 133, 170

Index

247

semi-empirical method, 27, 28

sensing, 45, 67, 125, 133, 134, 141, 171, 173, 174, 178

sensitivity, x, 3, 55, 67, 115, 123, 125, 174

sensitization, 107, 170

sensors, x, 175, 193

serum, 3, 45, 57, 60, 82, 83, 86, 87, 94, 97, 106, 109, 111, 112, 117, 118, 173

serum albumin, 3, 45, 57, 60, 82, 83, 86, 87, 94, 97, 106, 109, 112, 117, 118, 173

servers, 86, 102, 104

shape, 60, 64, 75, 83, 117, 129, 135, 139, 145, 159, 162, 163, 177, 220

showing, viii, x, 2, 22, 23, 25, 31, 32, 136, 164, 166, 169, 194, 200, 218

side chain, 31, 60, 111

silver, 55, 107, 133, 141, 164, 170, 176, 182, 183, 184, 187, 190, 191

simulation, 51, 163, 177

simulations, viii, 2, 5, 12, 22, 31, 35, 39, 48, 163

small molecule inhibitors, ix, 2, 3, 18

sodium, 128, 210, 219

software, 11, 12, 60

solar cells, 55, 107, 170

solubility, viii, 2, 11, 128, 216, 219

solvents, vii, x, 45, 46, 58, 76, 79, 174, 193, 194, 197, 198, 199, 200, 203, 204, 209, 213, 216, 221, 223, 234

sorption, 139, 152, 154

species, 11, 16, 20, 23, 26, 30, 41, 56, 65, 77, 79, 85, 89, 90, 91, 94, 95, 96, 97, 101, 103, 105, 106, 125, 132, 144, 160, 172, 203, 214, 215

spectrophotometry, 141, 142

spectroscopic properties, 126, 153, 191, 205

spectroscopy, x, 5, 105, 110, 123, 135, 136, 142, 146, 147, 148, 149, 164, 174, 185, 209

spine, viii, 2, 32

stability, 4, 12, 37, 68, 96, 134, 137, 138, 145, 153, 173, 194, 195, 197, 208

stabilization, viii, 2, 26, 34, 35, 36, 38, 39, 46, 117, 173

stabilization of the native insulin structure, viii, 2

stable complex, viii, 2, 30, 34, 70

stable complexes, viii, 2, 30, 34, 70

stoichiometry, ix, 54, 59, 94, 95, 109, 120

storage, vii, viii, 2, 4, 202

stress, 36, 50

substitution, 64, 68, 86, 111

substitutions, 68, 105

substrate, 125, 140

Sun, 19, 42, 177, 234, 235

surface area, 11, 12, 44, 48, 135, 139

surfactant, 111, 140, 219

surfactants, 56, 111, 203, 207, 219

suspensions, 135, 137

synthesis, ix, 54, 57, 136, 189, 195, 198, 201, 203

T

target, 29, 171, 172

tau, 4, 18, 26, 41, 42, 47, 50, 119

TDBC, 188, 210, 211, 212, 213, 214, 215, 216, 217, 218, 219, 220, 221

TEM, 22, 25, 135, 139, 140, 141, 150, 151, 174

temperature, 12, 56, 147, 158, 196, 204, 207, 212

therapeutic agents, 19, 34

thermal stability, 36, 196, 197, 202

thiacyanine dyes, 124, 125, 126, 133, 141, 142, 143, 164, 169, 174, 181

toxicity, 3, 42, 46, 194, 200, 201

trajectory, 35, 36

transformation, 19, 34, 60, 112

transmission, 22, 120, 135, 139

transmission electron microscopy, 22, 120, 139

transport, 44, 48

treatment, 41, 185
tumor, vii, 2, 3, 40, 55, 108

urea, 199, 201, 217
uv-vis absorbance spectra, 210, 217

viscosity, 195, 197, 202
visualization, 67, 114
volatility, 194, 195

W

water, viii, 2, 6, 10, 65, 78, 82, 114, 117, 128, 143, 163, 183, 188, 197, 200, 201, 207, 210, 212, 213, 216, 217, 219, 221, 234
wavelengths, 56, 69, 73, 74, 91, 97, 130, 133, 136, 148, 204
workers, 143, 198, 199, 200, 204, 208, 209, 211, 212, 213, 214, 215, 216, 217, 218, 220

Related Nova Publications

BIOCHEMISTRY LABORATORY MANUAL FOR UNDERGRADUATE STUDENTS

AUTHORS: Buthainah Al Bulushi, Raya Al-maliki, and Musthafa Mohamed Essa, Ph.D.

SERIES: Biochemistry Research Trends

BOOK DESCRIPTION: This laboratory manual has been designed for nutrition students for a better understanding of the lab assessments including biochemistry and food chemistry lab assessments. This manual includes both qualitative and quantitative analyses of some of the macro and micronutrients.

ONLINE BOOK ISBN: 978-1-53614-967-8
RETAIL PRICE: $0

AMYLASES: PROPERTIES, FUNCTIONS AND USES

EDITOR: Nikhil Adam

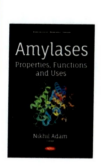

SERIES: Biochemistry Research Trends

BOOK DESCRIPTION: *Amylases: Properties, Functions and Uses* opens with an analysis of the methods commonly used for the immobilization of amylase on particles, the effect that the processes of adsorption and covalent immobilization have on the activity and stability of the enzyme, as well as on its stability and reusability.

SOFTCOVER ISBN: 978-1-53614-993-7
RETAIL PRICE: $82

To see a complete list of Nova publications, please visit our website at www.novapublishers.com

Related Nova Publications

HORSERADISH PEROXIDASE: STRUCTURE, FUNCTIONS AND APPLICATIONS

EDITOR: Maarten Uzun

SERIES: Biochemistry Research Trends

BOOK DESCRIPTION: In this compilation, the authors discuss the commercial source for the enzyme horseradish peroxidase, the tuberous roots of the horseradish plant which is native to the temperate regions of the world. Horseradish peroxidase is an oxidoreductase belonging to the highly ubiquitous group of peroxidases, indicating that this enzyme came into existence in the early stages of evolution and has been conserved thereafter.

SOFTCOVER ISBN: 978-1-53615-912-7
RETAIL PRICE: $95

HEMAGGLUTININS: STRUCTURES, FUNCTIONS AND MECHANISMS

EDITORS: Tzi Bun Ng, Jack Wong, Ryan Tse, Tak Fu Tse, and

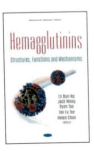

SERIES: Biochemistry Research Trends

BOOK DESCRIPTION: Hemagglutinins refers to glycoproteins which bring about agglutination of erythrocytes or hemagglutination. Hemagglutination can be used to identify surface antigens on erythrocytes (with known antibodies) and, hence, the blood type of an individual.

HARDCOVER ISBN: 978-1-53615-708-6
RETAIL PRICE: $230

To see a complete list of Nova publications, please visit our website at www.novapublishers.com